BRUNEI DARUSSALAM: ITS LOCATION IN BORNEO AND MAJOR RIVERS AND TOWNS

Kuala Belait, Tutong, Bandar Seri Begawan and Bangar are the administrative centres of the Belait, Tutong, Brunei/Muara and Temburong Districts, respectively

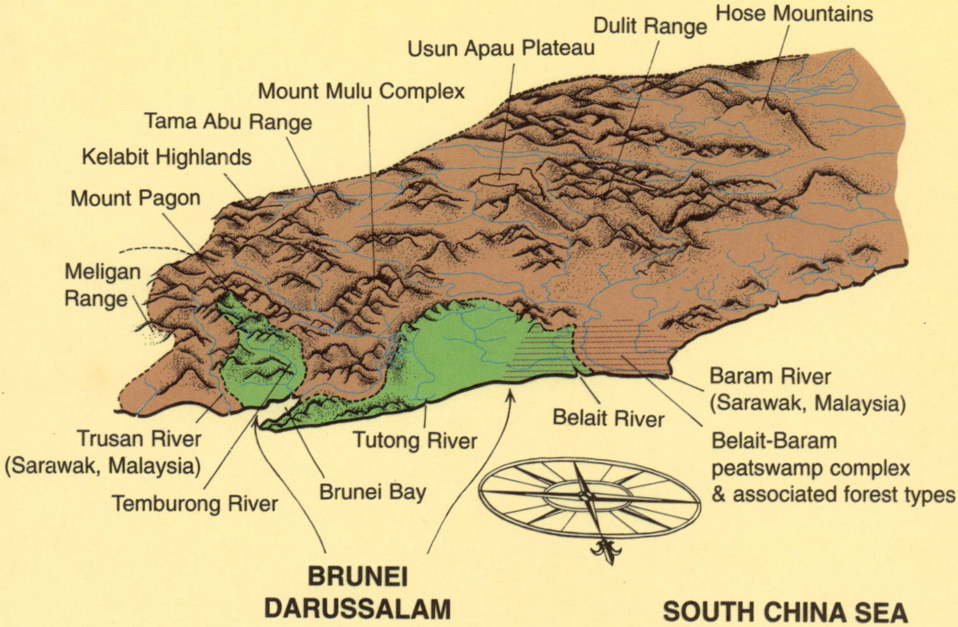

OBLIQUE AERIAL VIEW OF BRUNEI AND SURROUNDING AREAS FROM THE SOUTH CHINA SEA, SHOWING THE MAIN PHYSICAL FEATURES

In Brunei Forests

In Brunei Forests

An Introduction to the Plant Life of Brunei Darussalam

A Revised Edition

K.M. Wong

with watercolours by
C.L. Chan

Natural History Publications (Borneo) Sdn. Bhd.
Kota Kinabalu

in association with

Forestry Department
Ministry of Industry and Primary Resources
Brunei Darussalam

1997

Published by

Natural History Publications (Borneo) Sdn. Bhd.,
A913, 9th Floor, Wisma Merdeka,
P.O. Box 13908,
88846 Kota Kinabalu, Sabah, Malaysia.

in association with

Forestry Department,
Ministry of Industry and Primary Resources,
Brunei Darussalam.

First edition published 1990.
This revised 2nd edition published 1997.

IN BRUNEI FORESTS
An Introduction to the Plant Life of Brunei Darussalam
by K.M. Wong
with watercolours by C.L. Chan

Design and layout by C.L. Chan and Chua Kok Hian

Perpustakaan Negara Malaysia Cataloguing-in-Publication Data

Wong, K.M. (Khoon Meng)
In Brunei Forests: an introduction to the plant life of Brunei Darussalam / K.M. Wong ; with watercolour paintings by C.L. Chan. —A revised ed.
ISBN 983-812-012-X
1. Forests and forestry—Brunei. 2. Rain forests—Brunei. 3. Forest ecology—Brunei. 4. Forest management—Brunei. 5. Forest conservation—Brunei. 6. Botany. I. Chan, C.L. (Chew Lun). II. Title.
634.909595

Printed and bound in Malaysia

Contents

Illustrations

Foreword

There is no doubt that the tropical rain forest is one of the greatest treasures of our Earth. The great biodiversity of the South East Asian rain forests, and especially that of Borneo and Brunei Darussalam, has been widely recognised by scientists although there are very few books available to the general reader which give a summary of the uniqueness, usefulness and value of these forests. I strongly believe that the wonders of the rain forest should be translated into a book of the present form, where scientific knowledge is woven together and presented simply to provide a general understanding of the rich plant life in the rain forest.

The forests are not only a resource to be maintained for the many services and goods they provide. The species-rich tropical rain forests are also a "treasure chest" in which more and more useful materials and options will be discovered and which will be the basis of new and important industries. It will be necessary, however, to ensure that the pursuit of economic activity and development will not erode the forest cover to an extent where it is no longer effective as a moderating factor of the environment and as a natural conservatory for the thousands of plant and animal forms that are integral to the survival of the ecosystem. We need to recognise that industrial development without due regard for environmental conservation is potentially perilous.

Effective conservation and environmental stability can only be guaranteed by sound policies, which in turn require the support of science to provide the guidelines. Equally important is the incorporation of a basic appreciation for the environment and natural ecosystems in our general awareness. In this regard the present book is an important addition to our knowledge on the plant life and natural history of the rain forests of Brunei Darussalam. One cannot truly speak of sharing an international concern for the environment without first addressing the need for adequate conservation of natural systems within the country. Consequently I am whole-hearted in congratulating the Forestry Department in its commendable attitude to conservation which is elegantly expressed through this book, which I believe is an important introductory account of the plant life of Brunei Darussalam as well as an eloquent appraisal of the value of these species-rich forests.

HAJI ABDUL RAHMAN TAIB
(Pehin Orang Kaya Setia Pahlawan)
MINISTER OF INDUSTRY AND PRIMARY RESOURCES,
BRUNEI DARUSSALAM

Preface

In *Brunei Forests* is an attempt to present the fascination of the plant life of Brunei Darussalam, which is among the most diverse and interesting ever documented. It is difficult to write a book on a subject so involving as this and yet summarise the information in a way that could appeal to the general reader. The author has elected to use this style, which I strongly endorse, in the belief that there must be popular appeal and interest in the overall effort of conservation.

As the book makes clear, the wealth of plant life is both resource and heritage. Raising conservation consciousness, however, is not the sole intention of this account. The book explains the factors which have contributed to a high biodiversity in the forests of Brunei Darussalam and goes on to provide a panorama of the different types of forest and plant forms, and is therefore also an important basic reference. Selected examples of plant life and special aspects of biology are highlighted in short accounts accompanying the illustrations, reproduced from watercolour plates prepared by the illustrator. This provides an opportunity to turn to the different illustrations and glean some biological insight into the species or group of plants illustrated, in addition to the main text which is presented in four parts.

As is also explained by the book, the scientific documentation of tropical rain forests all over the world is far from complete. The botanical inventory of Brunei Darussalam, begun in 1989 as a special project to extend earlier information and conducted under the joint auspices of the Forestry Department, Brunei Darussalam, and the Royal Botanic Gardens, Kew, aims at a practical annotated checklist of Brunei plants and the accumulation of reference specimens and information related to the rich plant life of the country. This effort, essentially long-term in nature, will from time to time produce accounts of individual plant groups, vegetation types or forest products. The most crucial aspects will be the botanical field work and the subsequent verification of plant identifications. This is of basic importance because a knowledge of the basic content of the forest is required for refining guidelines useful to forest management and conservation. In the meantime, there is also a need for texts of general interest, so that the rich forest resources and heritage of Brunei Darussalam can be appreciated more widely. I believe this book, the first that is devoted to the plant life and forests of Brunei Darussalam, is an important step in the direction of scientific progress and sound conservation management.

HAJI MOHAMMAD YASSIN BIN AMPUAN SALLEH
DIRECTOR OF FORESTRY,
BRUNEI DARUSSALAM

Acknowledgements

The whole project of producing this book has received the constant encouragement and support of the Minister of Industry and Primary Resources, Brunei Darussalam, Pehin Orang Kaya Setia Pahlawan Dato Seri Setia Haji Abdul Rahman Taib, and the Permanent Secretary to the Ministry of Industry and Primary Resources in 1990, Dato Paduka Malai Ali bin Malai Haji Othman. Thanks are also extended to the present Permanent Secretary, Dato Paduka Awang Haji Ahmad Bin Pehin Orang Kaya Di Gadong Seri DiRaja Dato Laila Utama Awang Haji Mohd. Yussof, for facilitating the publication of the second edition.

Throughout, the progress in preparing the text has been facilitated by Brunei Darussalam's Director of Forestry, Haji Mohammad Yassin bin Ampuan Salleh and his successive Deputy Directors, Haji Hafneh bin Mohammad Salleh and Haji Abdul Rahman Bin Haji Chuchu. We are also grateful for various comments and suggestions that have improved the text, particularly in the present edition, from Dr Peter Becker of the Biology Department, Universiti Brunei Darussalam; Dr Christopher K. Morley of the Petroleum Geoscience Department, Universiti Brunei Darussalam; Dr John Dransfield of the Royal Botanic Gardens, Kew; and Mr Wong Tuck Meng of the Forestry Department, Brunei Darussalam. Mr Imran Yunus offered many useful criticisms for the revision and also kindly proofread the work. Mr Chua Kok Hian kindly assisted with the compilation of the indices. Assistance in the identification of plants was received from Mr Leonard Forman and Mr Barrie Blewett of the Royal Botanic Gardens, Kew, Dr P.S. Ashton of Harvard University, and Dr E.F. de Vogel of the Rijksherbarium, Netherlands. Ms Jolanda Huisman and Dr J.T. Weibes of the Museum of Natural History, Leiden, Netherlands, and Dr Gaden Robinson and Dr D. Agosti of the British Museum (Natural History) gave time to the identification of insects associated with the plants highlighted here. Ms Lee Su Win helped in sourcing literature related to a great number of topics reviewed during the writing of the text. The excellent illustrations used here are the work of the greatly talented Mr Chan Chew Lun who also freely gave information on orchids and insects. I am also grateful to Mr Yap Yun Hun for preparing the endpaper illustrations. Inspiration for writing many portions of the text came from field work carried out in the warm companionship of members of the Brunei Forestry Department's botanical team; Mr Niga Nangkat and Dayang Halimah Haji Morni maintained steady support during field and herbarium surveys, respectively.

For the original edition, the suggestions and editorial skills of Mr Nicholas Martland have been of immense value, and Mr Lo Wan Yin, Dayang Hasnah binti Haji Baki and Dayang Roslinah binti Haji Mohsin assisted with the production; the help of the Government Printing Department, Brunei Darussalam, and its Director Awang Ali bin Haji Hassan has been invaluable. Dr Kamariah Haji Abu Salim of the Biology Department, University of Brunei Darussalam, provided much assistance in the preparation of the second edition.

A Paradise of Plants

Brunei Darussalam—literally, "Brunei, Abode of Peace"—has some of the richest biological assemblages in the world, where an amazing diversity of life form and species continues to exist in rain forests that remain safe from destruction. This great biodiversity is a result of the interplay of different factors: notably, the location of Brunei in a region with a long history of land formation and speciation within a climate conducive to luxuriant plant growth, the intermingling of species from different land masses since ancient periods, and the existence of a large range of micro-environments within the complex framework of the forest to which different species may be adapted. Within this setting there is a paradise of plants.

Those fortunate enough to have been in the pristine rain forests of Brunei Darussalam have exulted over the experience. This is not unusual, when we realise that amid worldwide concern over large-scale decimation of tropical rain forest, some 80 percent of Brunei—or 4700 out of its total land area of 5765 square kilometres—remains under forest cover. Brunei is not a large country, but its overall wealth and rich heritage includes these rain forests, which are everywhere acknowledged as some of the richest ecosystems on earth.

This fortunate situation is a result of several factors. From the 1930s, the country has built up a system of forest reserves where timber-taking has been permitted over limited areas under strictly controlled licensing, and clear felling is not allowed. The wealth of the country, and a population that is largely urban and settled in the major towns along the coast have also removed any unfortunate pressure on these rich forest areas, and ecologically damaging intensive timber extraction has not been carried out. Today, Brunei remains one of only very few countries in the tropics where widespread forest clearing for timber exports or for shifting cultivation has not occurred, and more than 70 percent of its forests are still primary undisturbed rain forest.

For a country of its size, it is amazing that Brunei's forests contain some of the richest biological assemblages in the world. Seed plants (plants that produce seed unlike, say, ferns or mosses), which are the dominant plant group in any forest, are represented by an estimated 5,000 species, including more than 2,000 species of tree alone, in Brunei. The British Isles, about 54 times larger in land area, have 1430 species of seed plants. Peninsular Malaysia, about 23 times larger than Brunei in land area, has over 8000 species of seed plants. These are indications of an exceptionally rich flora. In well-studied groups of plants, the species counts equally reflect the great biodiversity found in Brunei. For dipterocarps—those members of the *damar* tree family which, from the viewpoints of their predominance in many forests and their timber value, are the most important group of trees in South East Asia—recent accounting listed 387 species for Malesia, the botanical region that includes the Philippines, Malaysia, Brunei, Indonesia and Papua New Guinea (Malesia is not to be confused with the political entity Malaysia; botanists have recognised Malesia as a coherent botanical region with a distinctive flora that is significantly different from that of the surrounding drier regions in the South East Asian–Australian area). This includes 268 species for the whole island of Borneo, and 154 for Peninsular Malaysia. In Brunei alone, in spite of the small land area, 171 species of dipterocarp have been documented: this is more than 60 percent of the species recorded for the botanically rich island of Borneo, and just under half of all species known from the Malesian region!

High plant diversity: an interplay of historical, climatic and evolutionary processes

Why is the flora in Brunei, or for that matter in the whole of the botanical region of Malesia, so rich? It may be said that this has resulted from an interplay of climatic, historical and biological processes. The constant humid climate of the tropics all year round encourages a luxuriance in plant life. There is no disruption by severe seasonal changes in temperature, sunshine or rainfall that restrict plant life in temperate climates. A long period of plant evolution in the region where Brunei and the island of Borneo are situated, since flowering plants appeared more than 160 million years ago, has also encouraged a profusion of species. In the course of such a long history, many species have evolved, some surviving only in particular sites where conditions are suitable, and others more widespread because of wider tolerance and more effective dispersal.

Within the rain forest, the structure of the forest itself is conducive to specialisation and thus, diversification of life forms and species. Within the framework built up by trees there is not only variation on the ground from one spot to another, but also variation upwards from ground to canopy. Within this framework there are many micro-environments to which species can adapt. As some plants grow independently upwards as trees, others are creepers, vines, herbs or epiphytes living on the branches and stems of other plants. As trees die and come tumbling down, they leave a break in the canopy of crowns, through which more light enters, thereby modifying the environment. Through the cycle of life and death, different micro-environments are created, at different times, in the forest.

Hence, the rain forest is a structurally and biologically complex system in comparison to other vegetation types, where the living space within is vertically extended and local environments change with time—it is here that "niche differentiation" is enhanced. In time, species adapt not only to conditions which are most conducive for their survival, but sometimes also to other species, such as the specific adaptation of certain parasitic plants like *Rafflesia* to certain species of host vines. Also, the competition among plants for resources may result in natural selection for different forms or populations that survive better in slightly different conditions (or, as some say, different parts of the environmental gradient). And when they are sufficiently isolated from one another and do not interbreed naturally, they become distinct species. Thus, in the *kapur* genus of dipterocarp trees, *Dryobalanops lanceolata* is common only on moist clay slopes of low hills, *Dryobalanops aromatica* and *Dryobalanops beccarii* occur on the more friable sandy soils, and *Dryobalanops rappa* is a tree of peat swamps and swampy sand terraces. Another example of this ecological differentiation is shown by the *tunjang pipit* palm (*Eugeissona minor*), common on the deeper clay and alluvial soils in Brunei's lowlands, in contrast to its relative, the *luba* (*Eugeissona utilis*), which occurs principally on the shallow sandy soils on steep ridgetops and slopes. Different species of the fern *Asplenium* are either terrestrial or grow on tree trunks and crowns. Even closely related species of plants in the same forest site, such as a number of closely related *Shorea* dipterocarp trees, may flower at different times, avoiding competition for insect pollinators.

In the Malesian region, the enriched variety of plant life has also come about through an intermingling of the floras of different regions. Scientists are still unravelling the events of the past, but it is likely that enrichment of the Malesian flora was facilitated when a major continental block of the ancient southern land mass Gondwanaland (the Sundaland block, which includes modern-day south Tibet, Burma, the Malay Peninsula and Sumatra) began detaching about 180 million years ago and drifted towards the southeastern margin of the northern continent called Laurasia. Other fragments of the southern continent included portions equivalent to west Borneo, east Borneo, west Sulawesi, east Sulawesi, the Banda and Timor islands, and portions of New Guinea. By about 90 million years ago, such fragments were above sea level and could have served as "stepping stones" for the dispersal of species in both directions. As late as 15 million years ago, the ancient Australian and New Guinean portions of Gondwanaland collided with southeast Asia, further increasing the chances for species dispersals. Thus, the Malay Archipelago is largely of southern origin but from very early on, the configuration of land fragments had allowed the dispersal of species from the north to mingle with those carried away from the south. Today, the flora of Malesia is estimated to include some 40,000 species of flowering plants, equivalent to at least 10 percent or more of the world's flora.

It appears also that historical climatic changes during glacial times of the Quaternary period (from about one million years ago) could have adversely affected the vegetation of Malesia as it had that of other tropical areas. During such periods of global cooling, when much water was locked up as ice, the sea level was as much as 180 metres lower than present and more seasonal climates were prevalent in the tropics. Many tropical rain forest areas contracted, accompanied by large-scale extinctions, and today some of the the richest floras exist in areas (called refugia) which have escaped the most adverse effects of such disruptions. It is possible that the northwestern part of Borneo was not as adversely affected as other portions of south east Asia, so that in such pockets or refugia, species richness had been largely maintained.

The flora of the northwestern part of Borneo including Brunei and much of Sarawak, Peninsular Malaysia and Sumatra share some similarities. When sea levels in the past were lower, the area that is at present the South China Sea separating Borneo from Peninsular Malaysia and Sumatra was for the most part exposed, much of it swamp and mangrove forests. These links also meant an exchange of species among the three areas, and today we can find species common to two or three of these localities but not occurring elsewhere. The Sealing Wax Palm (*Cyrtostachys renda*), today much cultivated as an ornamental, is a species found naturally in the peat swamps of Sumatra, Peninsular Malaysia and Borneo. The giant epiphytic screwpine, *Pandanus epiphyticus,* frequently seen in lowland riverine forests in Brunei, is known only from northwest Borneo and the southeastern part of Peninsular Malaysia. These are examples of a whole host of plants which link the flora of Borneo, Sumatra, and Peninsular Malaysia, on the Sunda continental shelf.

The botanical wealth that we now experience in the Malesian region is therefore a manifestation of a long history of speciation in an equable climate without severe disruptive events; of the intermingling of floras of different land masses since ancient times, and of the intrinsically complex and competitive environment within the rain forest itself. Thus there is a mixture of different floristic elements, some common to several places in the region, and others unique or special to just one area.

Barriers and corridors

The territory of Brunei has special geological and geographical features which may also engender a rich flora. The present area of Brunei, with the exception of central and southern Temburong, was dominated by a shallow sea from the middle Miocene to Pliocene epochs (approximately 16–1 million years ago). The sediments deposited during this time period formed extensive sandstone and shale beds several kilometres thick. During deposition, periodic phases of deformation folded, faulted and uplifted the sedimentary rocks, which were subsequently eroded. The result was the creation of three large, broad geosynclines (bowl-like structures) separated by narrow, more highly deformed anticlinal areas where the strata are arched upwards, forming prominent ridges. These geosynclines are the Belait geosyncline, the largest (including today's Belait, Tutong and Brunei-Muara districts); the Limbang syncline (separating the western and eastern portions of Brunei); and the Labu syncline (in the northern part of Temburong, the eastern portion of Brunei). The southern half of Temburong differs from the rest of Brunei, having folded and eroded lower-Miocene (older) sandstones and shales exposed. These rocks are harder (belonging to the same type of deposits that also form the Crocker Range in Sabah) and their erosion has left a complicated system of steep, sometimes almost knife-edged, ridges and valleys.

These synclinal basins are among several clustered together as part of a similar geological formation in this part of northwest Borneo, and therefore form a geologically distinct environment with its own characteristic sandy and clay soils. In addition, the Brunei area is somewhat surrounded by the more ancient Crocker, Meligan, Mulu and Tama Abu ranges and the Kelabit Highlands in the surrounding portions of Sabah and Sarawak, and the large Belait-Baram peat-swamp areas to the west. The peat swamps developed only some 6,000 years ago.

The Brunei area is thus something of an ecological enclave and biological assemblages occurring there are in a sense geographically and ecologically isolated. It is believed that biological isolation is a main factor in causing populations of species to diverge from their ancestral forms, and in this way new forms of life can arise in such pockets. The highlands that partly enclose the Brunei area may be effective geographical barriers for some groups of species, and the Belait-Baram peat swamps to the west are in their own way an ecological barrier for species that cannot survive in these swamps. Furthermore, the complicated pattern of steep ridges and valleys such as in the Temburong area may encourage localised ecological isolation, again important in encouraging the development of new forms and species. It may not be surprising, then, to find an especially rich and diverse flora with distinctive rarities in such an area. Examples include the bamboo *Temburongia simplex*, the sole species of an endemic genus occurring only in the valleys of the Temburong district, and the riverbank *keladi*, *Hapaline celatrix*, one of only two species of this genus found in Borneo and apparently also restricted to the Temburong area. *Pinanga yassinii*, an elegant forest palm, is known only from the Ulu Ingei hills of the Belait district.

On the other hand, mountain ranges which are geographical barriers can also be special corridors for the migration and intermingling of the specially adapted species of the mountains. Thus, the sandstone mountains of the Pagon and Retak ranges in the southern tip of Brunei's Temburong district are not only a refuge for montane species, but may also be important as midway points in the distribution of such species in the system of ranges including the Crocker, Meligan, Mulu and Tama Abu ranges in this part of Borneo. The tree *Prunus oocarpa* (Rosaceae) occurs on Mt Kinabalu at the northern end of the Crocker Range in Sabah, Mt Pagon in Brunei and the Kelabit Highlands in Sarawak, as does the montane bamboo *Kinabaluchloa nebulosa*. The large alliance of montane species of more widespread distribution in Borneo and which also occur in the high Temburong mountains includes the orchid *Entomophobia kinabaluensis*, the pitcher plant *Nepenthes lowii*, and a number of rhododendron species.

In Brunei today, we find the continued survival of exceedingly species-rich forests, a product of millions of years of biological and geological processes. The plant and animal life in these forests, of bewildering variety, and the special communities they form are the natural heritage of Brunei and, indeed, part of the great heritage of the Earth. The profusion of species is in particular readily observed—the diversity, the complexity and the luxuriance of the rain forest is reflected in this paradise of plants.

Plate 1

Shorea macrophylla

A Bornean "illipe nut" Plate 1

Well known in Brunei as the *kawang jantung*, this important dipterocarp tree frequently occurs on alluvial sites in lowland forest. Growing up to 30 metres high in the forest, it is occasionally planted in villages for its oil-rich winged fruits. The oil extract has been and is still used for cooking and flavouring rice throughout Borneo, as well as for fuelling torches and caulking boats in the past. Tens of thousands of tons of these nuts have been exported to Europe and Japan in recent years, where the oil is used as a substitute for cocoa butter in the confectionery industry and for the manufacture of cosmetics, candle and soap industries.

The Dusun people call this species *pitun* and the Ibans call it *engkabang*. The term *illipe* nut—derived from Tamil names for nuts of *Bassia* trees in south India—has been broadly applied to include the oil-yielding nuts of dipterocarps. Among the so-called illipe-nut dipterocarps, mostly called *kawang* in Brunei, *Shorea macrophylla* is the principal source of the fruit, although in all perhaps six species are important.

The fruits are collected in great quantities, their wings are removed and the nuts are usually first soaked in a river to increase their oil content before being kiln- or sun-dried. Dried fruits are crushed and steamed in a segment of bamboo or a tin, then the material is pressed in a mesh between wood blocks to extract the oil, which hardens within half an hour of collecting in bamboo tubes. In the hardened form, this can be stored.

There is great interest in *kawang* fruit harvests because these dipterocarps, like many others, do not flower or fruit every year. Such "mast fruiting" with intervals of several to many years has been interpreted as a habit evolved to allow large intervals between bumper fruit crops so that even predators, including insects, will be satiated, leaving enough uneaten seeds to produce the next generation. Certainly, deer, squirrels, wild boar and other animals flock where *kawang* trees fruit, and wild boar have been known to follow a wave of such fruiting. In longhouse communities it is an occasion of note, for not only will there be a bountiful harvest of illipe nut, but also forest animals will be common and adequately fattened in the area.

Plate 2

Aerides odorata

Fragrant gem among orchids Plate 2

Orchids form one of the most diverse families of flowering plants in the world. They occur all over the globe except at the poles, and are most abundant in the tropics. Estimates of the size of the family range from 17,500 to some 25,000 species. In the South East Asian region alone specialists think there may be 3,000 to 4,000 species, or even more.

Aerides orchids belong to the Sarcanthine subtribe of monopodial-growth orchids, which has perhaps 1,000 species. *Aerides odorata,* distributed from the Himalayan area through to the Celebes and wild in the lowland and hill forests of Brunei, is a relative of the Philippine *Aerides lawrenciae,* a valued and commonly cultivated ornamental. *Aerides odorata* is a hardy orchid growing on trees producing dense blooms of sweetly scented flowers lasting up to a week. Individual flowers have delicate lilac-pink petals and a yellow lip.

Among orchids, highly specialized floral adaptations for attracting and manipulating particular insects to effect cross-pollination have invoked great fascination, and involve a great variety of structural modification and colourful display in the flowers. *Aerides* orchids, for example, are only known to be pollinated by *Xylocopa* carpenter bees. Specialisation of pollinators to flowers is common knowledge among orchid enthusiasts and although the biological aspects of these remain insufficiently studied, it is believed to operate partly to avoid competition for pollinators among different orchid species.

Plate 3

Chelonistele amplissima

A Bornean mountain orchid — Plate 3

Chelonistele is a small genus of some one dozen species of orchid, most of which are endemic to Borneo. The exception is *Chelonistele sulphurea,* distributed in Borneo, southern Philippines, Sumatra, Java and Peninsular Malaysia.

The species pictured here, *Chelonistele amplissima,* is a Bornean endemic of montane forests which was documented for Brunei only in the late 1980s. It has the largest flowers of the genus, known only to be pollinated by a species of *Vespa* wasps. It is common on some mountains and represents many species of plants for which we have little or no knowledge regarding their uses or ecology. In terms of the horticultural industry, however, it has limited commercial value because it is not a hardy species in cultivation and the flowers do not last long, although at times flowers seem abundant. In general, though, the ornamental orchid industry is principally based on the production of hybrids which combine the best qualities—of beauty, fragrance and hardiness—from different stocks, species or even genera.

There are perhaps 2,500–3,000 species of orchids in Borneo, including many that remain to be scientifically identified. Wild populations, especially of rare species, are often vulnerable to destruction by indiscriminate collecting despite being illegal to collect without permits.

Plate 4 (notes overleaf)

Hosea lobbii

Hose's flower, a Bornean speciality Plate 4

Belonging to the verbena family, and related to the clerodendrons, the climber *Hosea lobbii* is abundant in many places along the lower meandering stretches of some rivers, and in freshwater alluvial swamps in Brunei and Sarawak. It is called *galap-galap* in Brunei, although that name is applied also to other woody climbers, such as *Mussaenda,* which have conspicuously coloured and enlarged floral parts or bracts.

Hosea is a genus entirely restricted to Borneo, representing a true speciality of the rich Bornean flora, and is known by only one species, *Hosea lobbii*. It was originally described as a species of *Clerodendron* in 1885, but later the great botanist Henry Ridley realised its special significance and named the genus for Charles Hose, a District Officer of Sarawak's Baram district and one of the first collectors of this plant. The American botanist E.D. Merrill, who worked on an enumeration of Bornean plants, once listed this as *Hoseanthus lobbii*, but this is not now accepted.

The flowers are pendulous and formed at the tips of side-branches. Each has a complement of pale yellow, fused petals and delicate pink stamens. Most conspicuous, however, are the bright orange leaves that occur near the flowering branch tips which catch the attention even from some distance away; in comparison, the normal leaves are green.

Sporadic in flowering habit, *Hosea* deserves further trial as an ornamental plant. It was cultivated for a time in the Botanic Gardens at Singapore before the Second World War.

The Rain Forest

The richest ecosystem on earth, the rain forest teems with life. There is not just one form of rain forest, but many. In Brunei, several types can be readily distinguished by the structural characters of the trees, but there are also differences in species composition, terrain, substrate, elevation and local climatic regimes where they occur. Within each formation there may be special communities or associations of species, and often the apparent burst of complexity and disarray that first strikes the eye belies any patterns there may be. The rain forest is dynamic with changes in the growth, death and distribution of plants, the last influenced by chance and modes of dispersal and establishment. Scientists are beginning to document the biodiversity of forests, a basic step in learning to better manage them for the numerous products and services they provide.

There have been rumours for a long time of the existence of a magnificent forest in Brunei — wrote C. Hummel, Deputy Conservator of Forests of the Federated Malay States in 1914, when he was despatched to ascertain the validity of this statement. Hummel was overwhelmed by the large number of trees unfamiliar to him when he arrived in Brunei, and remarked: "To identify and describe these trees in full would probably take more than a year..." That Hummel was overly optimistic was expressed in a statement by Bertram Smythies, State Forest Officer in Brunei in 1956, that "...after *twenty-one* years of fairly intensive study we still have no name for many of our important dipterocarps." In 1957 Peter Ashton arrived as Forest Botanist to begin a special study of the dipterocarp trees. Supported by the intensive collecting of specimens before and during his three years of field studies in Brunei, Ashton finally prepared his "Manual of the Dipterocarp Trees of Brunei State," published in 1964. He recorded 154 species for Brunei. Ashton extended his studies on dipterocarps to cover the entire botanical area of Malesia (including Malaysia, Indonesia, the Philippines and Papua New Guinea), published as a special account nearly twenty years later, in 1982. Since then, even more has become known about the distribution of dipterocarps and today 171 species have been recorded for Brunei. Only as recently as 1990 did the Forestry Department's botanical team discover the rare *Dipterocarpus lamellatus* in Brunei; this is a large tree previously documented only from Labuan (where it is now extinct) and Beaufort in Sabah. Had Hummel realised the full implication of his observations in Brunei in 1914, he might have issued a more dramatic statement.

This is not surprising, for the tropical rain forest is the richest ecosystem on earth. What exactly is tropical rain forest and, aside from its overall great biodiversity, what variations of it are there? In general, this term refers to the so-called evergreen forests of the humid tropics, and although the forest as a whole is not deciduous, some species, such as the gigantic *menggaris* tree, can have bare crowns for a short time when a completely new flush of foliage develops. The three major blocks of tropical rain forest exist in South America, Africa and the Far East; the last includes the botanical region of Malesia which has some of the most species-rich communities ever documented. Only in the rain forests of the Far East are trees of the Dipterocarpaceae (i.e., the dipterocarp family) so noticeable in terms of abundance and variety. In the other rain forest blocks, other families of trees are significant. In general, however, the most extensive tracts of rain forest are those developed on well-drained ground, where the main structure of the forest is formed by trees of different species and families reaching to different levels from the forest floor, accompanied by a commonness of lianes (woody climbers) and other life forms including epiphytes (plants growing on other plants) and forest-floor herbs.

Rain forest growth habits

There are still other features that are typical of most rain forests. The most striking must be the preponderance of tall, columnar tree trunks which, in its grandest display, imparts a majestic character to these forests. After one becomes more familiar with being in such a forest, it becomes noticeable that many trees have buttresses that appear to help prop up the large trunks, an uncommon feature away from the tropics.

These buttresses can be small brace-like structures at the base of the trunk, or tall, spreading plank-like supports to more than six metres tall on the biggest trees. The occasional uprooted tree in the forest reveals yet another characteristic feature which is perhaps more easily noticed beside a road cutting through a forested slope: the roots of most trees do not reach very deep into the ground. In fact, most roots normally do not penetrate more than half a metre deep in the rain forest, and many a large uprooted tree appears to have a plate-like root system, with masses of roots running laterally and never deeply downward. This may be related to the inability of roots to respire efficiently at greater depths where oxygen is insufficient, and to the abundance of nutrients at the most superficial layers of the forest floor where organic matter is actively broken down by fungi, bacteria and small animals.

Also fascinating for the newcomer to a tropical rain forest is the frequency with which flowers or fruits can be found growing out of tree trunks (a phenomenon called cauliflory) or branches (ramiflory), instead of being always borne at the shoot tips or among the foliage. This can be commonly seen in some of the most popular fruit trees in South East Asia, including the *durian* trees (various species of the genus *Durio,* of which the species *Durio zibethinus* is the most frequently planted), the *tampoi* (*Baccaurea macrocarpa*), the *rambai* (*Baccaurea motleyana*) and the *langsat* (*Lansium domesticum*). This cauliflorous habit is most commonly seen in the lowlands where perhaps the example of the *durian* must be the most spectacular; different species of *durian* develop fruits on the trunk or branches, or both, and the massive prickly fruits (within which are borne the seeds enveloped in an edible, creamy flesh) should be the most unlikely things that are suspended from branches or trunks by their comparatively slender stalks. In Brunei, the *durian kura-kura* ("tortoise durian") bears its prickly fruits right at the base of the tree trunk, at ground level. It must be noted that this kind of flowering or fruiting is not restricted to trees, and occurs also in some woody climbers.

Rain forests in the tropics exist as different formations or variants. Which formation will predominate is influenced by such factors as whether the soil is well drained or the water table is high, the type of soil in an area, the distance inland and the elevation. In Brunei at least eight forest formations are easily recognised.

The Mixed Dipterocarp Forests

The most extensive formation is the lowland evergreen rain forest, often referred to in the western part of Malesia (including Brunei) as mixed dipterocarp forest by virtue of the large variety of dipterocarps which dominate the forest with their stature. This is the richest formation in terms of species diversity and the abundance of different life forms that can be found on well drained soil from sea level to an elevation of about 1200 metres. The forest is lofty, with the biggest trees, called emergents, protruding above the general canopy of crowns and reaching fifty metres or more in height. In its structural grandeur, lowland evergreen rain forest has no match. Yet, all over the tropics, these forests have been felled at alarming rates for their timber content.

Underneath the canopy of this forest is a profusion of trees of different sizes and kinds, and at different stages of development. Herbs are more abundant only in moister spots or where a gap in the canopy of crowns allows more light through to an otherwise well-shaded, and sometimes even dark, forest floor. Within the forest on a clear day, the atmosphere can be almost uncanny: there seems to be little in motion except when the highest branches sway in a breeze or when the occasional animal appears, and the silence is broken only by rustling foliage and the rhythmic calls of birds or insects. However, this is only a superficial impression, as much of the "action" is not easily perceived or is easily missed. Larger animals remain shy of the presence of the intruding observer, and the busy trails of ants and termites, ever present in the forest, are usually bypassed with little curiosity. Changes in plant growth can only be noticed with repeated observations at the same site and trees, which can live as long as half a century or more, have a different time frame of development compared to animals and man. Perhaps the best clues, at any one moment, to the dynamic nature of forest growth are the seedlings and sprouts, representing a renewal of plant life; the range of different tree sizes, representing the entire spectrum of the growth process; the dead trees and fallen boughs, which bespeak a natural end to the life of individuals; and the frequent untidy tangles of lianas which bind together tree trunks and crowns and which could only have been carried upwards by the trees on which they have successfully ascended in the initial phases of growth.

The rain forest is in a state of dynamism. Plants grow, reproduce and die. The organic matter, whether from the smallest leaf or the largest trunk, is broken down by microbes, fungi and animals, and the nutrients released through this process of decomposition are rapidly taken up by living roots. Animals consume plant parts and, when they die, decompose in the same way and this organic matter too is returned to the soil. There is a continuous recycling of nutrients in the forest system. Seeds and fruits drop to the forest floor where they rot, are eaten or germinate; or they may be dispersed by wind or animals to new sites. An emergent tree crashing down creates a light-filled gap that is rapidly colonised by seedlings and vines that grow vigorously under the enhanced light conditions. Such gaps initiate new growth and are an important feature of the forest growth cycle.

Mixed dipterocarp forests are sometimes categorised as lowland (roughly below 300 metres altitude) or hill types, but there is not any sharp transition, although at higher elevations the kinds of species present may differ. Even at low elevations there are differences in species composition which may subtly reflect variation in terrain, aspect and soil type. In Brunei, the mixed dipterocarp forest on hill ridges often carry larger trees with bigger crowns. Several illipe-nut dipterocarps, especially the *kawang jantung* (*Shorea macrophylla*), and the Borneo ironwood, or *belian* (*Eusideroxylon zwageri*) may be common in patches fringing occasionally inundated alluvial forests or along riverbanks. The species diversity of different sites may be related to the amount of nutrient elements, such as phosphorus and magnesium, available to plants. Besides mixed dipterocarp forests, there are four other significant dry-land vegetation formations in Brunei. These are the fascinating lower and upper montane forests above 1200–1300 metres, and the formations of *kerangas* and beach vegetation in the lowlands.

Beach vegetation

Beach vegetation in Brunei typically includes an association of herbs and creepers on the sandy beach and a narrow belt of trees and shrubs behind this. The most conspicuous trees include the *ru* (*Casuarina equisetifolia*), the *bintangor laut* (*Calophyllum inophyllum*) and the coppery-leafed *nyatoh laut* (*Planchonella obovata*). The shrubby

Scaevola sericea (taccada or Malayan rice-paper plant), with thick, pale green leaves and small, white flowers each in the form of a tube slit down one side, frequently forms a fringe along the landward margin of the beach. This vegetation has adapted to the salt spray from the sea, exposure to the sun and the wind, and the nutrient-poor substrate.

Kerangas or tropical heath forest

Kerangas is a term used by the Iban people for forested land which is not suitable for hill-rice growing, excluding swamp forest. Most *kerangas* vegetation resembles a forest of small trees, developed on a thick layer of nutrient-poor, white sand in the form of low ridges or flat terraces. These ancient sea beaches were left by a drop in the sea level about a million years ago and are today highly conspicuous just behind much of the Brunei coastline. On these severely leached sands, the vegetation has far fewer species than mixed dipterocarp forest and has a lower canopy without many truly big trees. In *kerangas*, streams of tea-coloured water, stained with humic acids, run through the white sands and peculiar stands of pole-sized trees, where the ground is frequently lightly covered with mossy patches. In sparser stands many herbs that grow well in bright conditions are found on and around the base of trees, and slender, wiry climbing plants are common, some presenting blooms that seem oblivious to the intense heat of day. In some places curious life forms abound. There are the insectivorous sundews (*Drosera*) which possess tiny rosettes of leaves bearing glandular hairs, with which unsuspecting small insects form a fatal bond. Pitcher plants (*Nepenthes*), with their leaf tips drawn out into expanded, highly ornate pitchers in which insects drown and are digested. Equally unusual are the epiphytic shrubs *Hydnophytum* and *Myrmecodia,* which attach to the trunks and branches of trees, their stems basally inflated into an enormous lump within which is a maze of tunnels inhabited by fierce black ants.

Different forms of *kerangas* are found as islands amid peat swamp forests and on sandstone ridges further inland from the coast. On the deep white-sand remnant terraces surrounded by peat swamps a taller *kerangas* develops, where the trees are bigger and where the majestic conifer *Agathis borneensis* (in Brunei called the *tulong* tree, and elsewhere, *damar minyak*) may grow abundantly. This conifer is related to the famed Kauri pines of New Zealand. *Tulong* trees are extremely rare in the lowlands, and such *kerangas* islands are very special habitats. On many sandstone ridges, the weathered sands support *kerangas* rich in big trees of the dipterocarps *Dryobalanops aromatica* and *Dryobalanops beccarii,* which form beautiful crowns resembling giant cauliflowers. These dipterocarps yield a crystalline camphor which was traded by the Arabs in the 6th century and is mentioned in the writings of Marco Polo in 1299.

Montane forests

Frequently shrouded in cloud and mist, montane forests begin to be distinct around 700–900 metres in the highlands in Brunei's Temburong district. There is no sudden transformation from the upper reaches of mixed dipterocarp forest, but a gradual change in the character of the forest. In lower montane forest, most trees attain heights of only 15–25 metres, and in general are distinctly smaller than the giants of the mixed dipterocarp forest lower down. Around some peaks the mountain *tulong* (*Agathis dammara*), like its sister species on the white-sand islands in lowland peat swamps, congregate as tall imposing individuals.

Still higher up on the mountains, above about 1300 metres, upper montane forest begins to develop, with stands usually only 10–15 metres tall, and composed of even thinner, frequently gnarled trees. Many plants here have noticeably thick and small leaves. The moisture-laden clouds that drift through these forests encourage a profusion of mosses, liverworts and ferns which drape the trunks and branches of the trees. This Elfin Woodland or Moss Forest, as it is sometimes referred to, becomes even more dwarfed at its uppermost reaches, where only a shrublike cover about 1.5 metres high exists. In these forests, orchids and rhododendrons abound, transforming their habitat into a garden when they bloom. Different species of pitcher plants thrive, some intertwined with the shrubbery, and others in the crowns of trees or snugly nestled in thick pads of mosses that carpet the forest floor in many places.

These dwarf forests are different, developing on the highest peaks and ridges up to Pagon, at 1850 metres, Brunei's highest point. In the mountains, temperatures are lower, as there is an average fall of about

7°C for every thousand metres' increase in elevation, so that plant growth and the decomposition of organic matter proceed more slowly. The slow plant growth is perhaps accentuated by thin, nutrient-poor soils developing on ridges and peaks, and the low decomposition rates by the tough texture of many leaves, which make them more hardy while alive but also more difficult to decompose when dead. The result is a dwarfing of trees and the accumulation of frequently thick layers of slowly decomposing organic matter or peat.

Mangrove forest

Few people have not heard of mangrove, one of the three principal swamp vegetation types in the tropics. Mangrove, sometimes referred to by the Portuguese term *mangal* or the Malay word *bakau,* is a distinctive tree-dominated community built up on the seaward margin of the land and along estuarine habitats, under the influence of the tides. Brunei's mangroves are mainly found around the Brunei Bay, into which many rivers empty and where there are many tidal islands, and at the mouth of the Tutong River. Still largely intact, these forests have in the past been exploited for wood used in making charcoal and for firewood and cutch, a dye obtained from trees and used in the tanning industry. Today harvesting of mangrove poles is allowed only in specially designated areas and the value of mangroves in coastal protection and wildlife conservation, and in the maintenance of forestry resources, is fully appreciated.

Mangrove forests are important in being able to stabilize mud flats in estuarine and coastal areas. Mangrove trees are well adapted for this with their supportive stilt roots, buttresses, strong platforms of spreading roots and even pneumatophores (breathing roots) that extend above the poorly oxygenated mud. *Bakau* trees (*Rhizophora*) are propped up by large stilt roots, while groves of *api-api* trees (*Avicennia*) send out long lateral roots on which develop vertical peg-like breathing roots, stabilizing newly piled up soft mud. Mangrove plants have developed a specialised mode of life that adapts them to sea water and to tidal mud which is, needless to say, a soft, unstable substrate. All species here are salt-tolerant to some degree and many plants have thick leaves which resist drying out easily. Some leaves have water storage tissues as well.

Bakau trees are known to have more erectly held leaves in exposed parts of the crown, seemingly to reduce the dessicating effects of the sun.

The muddy estuarine waters and mangrove sediments are rich in nutrients, and support the breeding of many types of fish, prawn, crab, clam, oyster and snail. Disturbance to mangroves will obviously affect the productivity of coastal fisheries. Mangroves, like other forests, support a range of animal life, including otters and the protected proboscis monkey. Birds such as eagles, egrets and kingfishers are associated with mangroves and the food-rich mudflats are used by a great many species of coastal birds, many of which are annual migrants from their breeding grounds as far away as Siberia, escaping the northern winter. On flights over distances as much as 10,000 kilometers to Australia, the mangroves of South East Asia are an important feeding and resting place for these birds.

Peat swamp forests

As the mangroves continue the seaward advance of estuarine plains, they leave behind an environment increasingly remote from tidal influence, where debris is gradually accumulated and drainage is poor. In such places the plant matter fails to decompose properly because the salt and sulphide content of the clays, built up under seawater conditions in the past, is toxic to micro-organisms which effect the decomposition processes. Here mangrove is replaced and peat begins to form, the basis of another wet-land forest formation. Like mangroves, peat swamp forests, which cover as much as 15 percent of the land area of Brunei and are an important forest type in the lowlands of south and west Borneo generally, have their own peculiar characteristics.

Peat swamps may develop in small river valleys but are most extensive in the form of coastal peat basins built up around estuarine plains. A single peat swamp may stretch to more than 50 kilometres across, and the peat can be as deep as 20 metres. The water table in a peat swamp is high but typically below the surface. The only source of incoming water is rain, and there may be occasional flooding or drying out of the surface. Peat swamp forests are inhabited by characteristic plant communities tolerant of a stressful environment represented by the mineral-poor and acidic nature of the substrate; the shortage of surface water available to the plants during dry periods; the poorly aerated conditions during waterlogging which prevent efficient water uptake by plants; and toxic substances in the highly acidic ground water. Peat swamps have about 150 tree species in common with *kerangas*, where plants have to cope with a similarly stressful environment.

One of the most fascinating peat swamps is in Badas in Brunei's Belait District. The swamp possesses a biconvex shape in cross-section and the sequence of changes in vegetation from the margin to the centre, where the peat is normally deepest, is readily observed. The lens-shaped structure is due to the manner in which peat accumulates and due to the deposition of alluvium on the margins by rivers, as the coastline advances and the oldest parts of the swamp are left further inland. The accumulation of peat as a shallow dome at first then graduates into a wide bog plain gently sloping outwards from the centre and with steep sides. The catena, or sequence of forest types from the periphery of the swamp to its centre is related to the characteristics of the peat dome. A mixed swamp forest, structurally similar to dry-land lowland evergreen rain forest, occurs in an outermost zone, and gradually transforms into a forest dominated by scattered large trees of one species: the dipterocarp *Shorea albida,* locally called *alan,* and which is endemic to Borneo, ranging from the south through Sarawak until the Tutong River in Brunei. The *Shorea albida* trees in this forest reach 60 metres high and have large, stag-headed crowns, mostly hollow stems and big buttresses and are called *alan batu* (*batu* in Malay meaning rock, in reference to the denser wood of this form). The forest changes into *alan bunga* forest further inward, where the *Shorea albida* has a lighter wood, no noticeable buttressing, and forms an even canopy 45–60 metres high. *Alan bunga* forest resembles a plantation by virtue of the sole dominance of the species as the canopy tree. From the air, *alan* is very noticeable by its characteristic, grey-green crowns.

Still further inward the vegetation becomes shorter, only about 25–35 metres high, where *Shorea albida* is dominant, or co-dominant with other species—this forest is called *padang alan* ("a field of" *alan* trees). Typically, a well developed peat swamp is characterised by even greater stunting of the forest towards the centre, where other species become more dominant. The most severely dwarfed forest at the centre of a swamp is only about 15 metres tall, consisting of stunted *keruntum* (*Combretocarpus*) trees as the dominant. Not every swamp has the entire sequence of forests, with some of the most dwarfed communities being absent. Important inorganic nutrients, such as nitrogen, phosphorus and potassium, become increasingly scarce towards the centre of the swamp. As in the case of montane forests, stunting of the vegetation is related to increasing infertility and peatiness, which reflect poor decomposition of dead organic matter and thus a less efficient recycling of nutrients in the system.

Within peat swamp forest are other curiosities of nature. The feathery-leafed Sealing Wax Palm (*Cyrtostachys renda*) with its brilliant red leaf bases (much sought after in ornamental horticulture), lives among taller communities of the peat swamp. On big trees of *Shorea albida,* the magnificent climbing orchid *Bulbophyllum beccarii* attaches its roots to the trunk as it spirals its way upwards. Its thick, dark green leaves can be as long as half a metre and the purplish blossoms, very rarely seen, are held on a main stalk almost ten centimetres long. The foul, carrion-like smell emitted by the flowers are thought to attract flies as pollinators. Especially common among the more stunted forests near the centre of the swamp and in patches of the taller peat-swamp communities, there are several types of plants that form an association with ants. The ground shrub *Clerodendron fistulosum* has ants in its hollow stem segments, and the large, conspicuous epiphytes *Hydnophytum formicaria* and *Myrmecodia tuberosa* develop inflated rhizomes riddled with ant-inhabited tunnels, mentioned earlier for *kerangas* forests. Some ferns participate in forming a specialised relationship with ants, called myrmecophytism, and two ferns in particular—*Lecanopteris sinuosa* and *Lecanopteris crustacea*—have hollow rhizomes in which the ants live. The detritus carried into the rhizomes or left there by ants are useful to the plant, which sends fine roots into the recesses of its rhizomes to reach this source of nutrition. It is thought that nutrients could also be directly absorbed through the inner lining of the rhizome. Four or five species of pitcher plants live in these peat swamps, but the most common may be the spectacular *Nepenthes bicalcarata,* with pitchers bearing two dangerous-looking curved fangs at the base of their lids. Biologists suggest that ant-associations and animal-trapping mechanisms are ways through which plants can increase their nutrition in nutrient-poor habitats such as peat swamps and *kerangas*.

Other features in peat-swamp plants are also comparable with those of other habitats. In some mixed swamp forests, knee-shaped roots emerge in profusion around the trunks of such trees as *Calophyllum obliquinervium* (of the Mangosteen family) and *Dyera lowii* (a relative of the *jelutong* tree, *Dyera costulata,* sometimes tapped for its white exudate used in the chewing gum industry) in the same way as those of the mangrove *lenggadai* trees (*Bruguiera*). These are specialised breathing roots that develop in poorly aerated substrates such as waterlogged peat or mangrove mud.

The complete dominance of one or a few species of tree in stressful habitats is seen in many peat swamps, mangroves and secondary forest on dry land. The peat-swamp species *Shorea albida* (*alan*), *Dryobalanops rappa* (*kapur paya,* another dipterocarp, related to the Borneo Camphorwood), *Combretocarpus rotundatus* (*keruntum,* usually considered a member of the *bakau* family, Rhizophoraceae) and *Campnosperma coriaceum* (*terentang,* of the Mango family) can individually form almost pure stands, as do several species of mangrove tree (notably *Rhizophora,* the *bakau*) and secondary forest (notably species of *Macaranga,* of the Rubber-tree family, locally called *sedaman*). How do the extremely stressful conditions in such habitats invoke the establishment of pure stands of individual species? We know that as more stressful conditions are met with towards the centre of a peat swamp there are fewer species growing. In a more general way, the flora of lowland mixed dipterocarp forest is incredibly richer than that of, say, a mangrove forest (where saline conditions and loose mud present insurmountable problems to the majority of plants) or *kerangas* (where the sandy soils are highly impoverished in nutrients). Where environments are more stressful, fewer species can adapt and, among these, several have become highly successful and aggressive in the way they establish themselves as pure stands to the exclusion of other species. All these species certainly do not achieve their dominance through the same means. For instance, *terentang* trees grow fast and seed regularly, whereas *kapur paya* grows more slowly and seeds only once in several years. Even more extreme is *alan,* which is known to have intervals of up to several decades between seedings. We simply do not know enough.

Freshwater swamps and *empran*

In contrast to peat swamp forest, freshwater swamp forest develops on fertile, alluvial soil that is regularly or periodically flooded over, usually beside slow-moving stretches of rivers. Though some plants are especially common in such forests, in general this formation resembles an extension of the species-rich lowland mixed dipterocarp forests. There are, however, other communities which thrive on some periodically inundated sites called *empran* by the Iban people. In its broadest sense *empran* includes open fields principally consisting of the *purun* sedge (*Lepironia articulata*), developing in broad seasonally water-filled alluvial valleys known in Brunei as *luagan,* as well as the tree-dominated freshwater swamp forests where several illipe-nut dipterocarps may also be common. Of these, the sedge-filled luagans provide a captivating contrast to rain forest. When the water level is high the entire valley resembles a lake, and during drier periods the thick growth of sedges present the appearance of a wind-blown field of tall grasses. Such visually striking transformations emphasize that even in the humid tropics seasonal climatic changes are ever present, as is also shown by the deciduousness of some species of rain-forest trees during drier months at the beginning of the year.

These rain forest formations and vegetation types reflect an amazing spectrum of plant communities over a large range of habitat conditions. Each has its own special characteristics, and all are an expression of the immense biodiversity which surrounds us. The explosion of complexity in the rain forest has patterns and processes which are recognizable but which require more detailed study. The task of protecting sufficient areas of the different formations and understanding the diversity of life in the various natural communities all over the tropics is one of great urgency. If we are to continue to benefit from these wonders of nature surely the greatest folly will be to allow the destruction of these forests before we even realise what treasures there are within.

Plate 5

Phalaenopsis cornu-cervi

Lowland forest orchid Plate 5

Phalaenopsis cornu-cervi occurs as isolated individuals attached to trees in lowland mixed dipterocarp forest, and is distributed from northern India, through Sumatra and Java, to Borneo. The brown-blotched, pale yellow flowers are borne, usually one at a time, along a gradually elongating inflorescence stalk which, after many flowers, develops to be a long stalk that can exceed sixty centimetres in length. Such a long inflorescence, carrying a fresh flower at its tip and tightly clasped by two rows of flattened, alternating old flower bracts just below this, can continue growing over a year and eventually protrudes laterally or hangs down from the leafy shoot, giving the plant a very characteristic appearance. The whole plant, with a few such long inflorescences, is a picture of elegance. The Latin epithet *cornu-cervi* means "deer horns", in reference to the forked appendages on the lip petal of the flower.

Most *Phalaenopsis* species have hardy growth and produce pretty, fragrant and long-lasting blooms. Commonly cultivated relatives of this species include *Phalaenopsis amabilis, Phalaenopsis gigantea, Phalaenopsis pantheriana, Phalaenopsis sumatrana* and *Phalaenopsis violacea.*

Phalaenopsis, as well as *Arachnis* and *Vanda,* are commercially important genera of orchids belonging to the Sarcanthine subtribe, of which numerous hybrids between different genera and species have been created by hand-pollination. Most hybridization work has emphasized larger blooms and attractive colours but the hybrids are mostly sterile, unable to produce seed. While this is of relative unimportance to the orchid grower, in nature many different species remain distinct through reproductive incompatibility, so that populations of two species cannot interbreed effectively. Thus, even when pollen from one species is used to pollinate flowers of another, the resulting hybrid offspring may be sterile.

Plate 6

Pandanus odoratissimus

The littoral screw pine — Plate 6

The screw pine, so called because the stems bear sword-like leaves in three ranks that spiral upwards as a helix, are estimated to include 500–1,000 species found in the Old World tropics. The common seaside screw pine of South East Asia, pictured here, is a usual component of beach vegetation and is typical of many species of the genus (*Pandanus*) which develop a tree-like habit, growing to several metres tall and sending out stout, tough prop roots from the trunk. These screw pines, as in the case of palm trees, have a ring of actively dividing cells around the base of the shoot tip which contributes to height and diameter growth, so that a seedling of narrow stem diameter produces a crown of increasing diameter as the plant grows taller.

The ripening fruit-head, which consists of numerous fused clusters of bright orange fruits set closely together on a large, deep red base, is probably the most bizarre part of this species. The heavy fruit-heads, superficially resembling pineapples, dangle from the tips of branches and disintegrate as they ripen. The individual fruit portions float, and are dispersed by sea water. In other species the fleshy-walled, big fruits may be dispersed by animals. Some New Guinean tribes use the soft inner parts of the fruit as food, either fresh or roasted, and sometimes reference is made to these species as "bread fruit".

In Brunei, the stemmed, tree-like screw pines are called *pandan* or *gadur*, of which several species yield a "cabbage" (the tender shoot tip) eaten as a vegetable. In contrast, stemless screw pines are called *mengkuang* if they reach sizeable proportions, and leaf strips of some of these are dried and woven into mats and baskets, as well as carrying bags, belts and rope.

Another relatively common screw pine in Brunei is a species that grows in the crowns of trees, appropriately named *Pandanus epiphyticus*. This species is common to the western rim of Borneo and the southeastern flank of Peninsular Malaysia, and is part of the botanical evidence that the two territories were once linked by forested land enabling the spread of species. In Brunei, *Pandanus epiphyticus* is common in riverine forest and often adorns the crowns of riverside trees, particularly those of the grand *ensurai* (*Dipterocarpus oblongifolius*) trees growing along the middle sections of Brunei's larger rivers.

Fagraea cuspidata

A *belukar* tree Plate 7

Called *kayu pitundok* in Brunei, this species is usually a small tree often found as a component of young secondary forest (or *belukar* in Malay) on cleared or exposed sites. The white, bell-shaped flowers are pretty, and are borne in hanging clusters from the branch tips. Mature trees are frequently in flower over many months, each bloom lasting two or three days. This, and the crown habit that resembles a coffee bush, makes it a likely candidate for an ornamental tree. As is characteristic of belukar species, the *kayu pitundok* grows at a much faster rate than primary forest trees, and rarely persists in the taller older forest that gradually replaces *belukar.*

The white flowers of the *kayu pitundok* are most likely visited and pollinated by nocturnal moths. Sticky dot glands are conspicuous on the young leaves and produce a sugary substance that attracts black ants to the shoots. These ants are fierce and bite when disturbed, and may play a role in protecting the young shoots of this tree.

The *kayu pitundok* is cousin to the well-known *tembusu* tree (*Fagraea fragrans*) and both are commonly classified by botanists as belonging to the family Loganiaceae, sometimes considered to also include the genus *Strychnos*. Wild strychnine vines (*Strychnos*) occur in Brunei's forests, and some species are generally known to contain the poisonous strychnine. This potent poison, however, is not found in *Fagraea* trees. In Borneo, more than 40 species of *Fagraea* occur, including trees, epiphytes and climbers, and among which are a number of very rare ones.

Plate 7

Agathis borneensis

Majestic conifer of the giant podzols — Plate 8

Known in Brunei as *tulong* (Malay and Dusun languages), *Agathis borneensis* grows in dense stands in a form of *kerangas* forest developing on deep white-sand terraces (called "giant podzols") in the midst of peat swamp, and occasionally on sandstone ridges and hillsides. The majestic, columnar trunks of these trees reach to thirty metres or more in height and are among the largest trees where they occur.

Agathis timber is well known and is classified in botany and in trade as a softwood, a term used to refer to wood where a type of water-conducting tissue called "vessels" is absent, in contrast to the wood of most flowering plants. In addition, *Agathis* trees—like other conifers—produce seeds that are not in any kind of specialised fruit structure, but which instead are attached to woody scales (inset, bottom left) arranged into cones. These are some of the characteristics that set aside the conifers as plants with a more ancient history, and which have retained more primitive traits, than the flowering plants.

Agathis borneensis, like other conifers, are wind-pollinated and its seed is wind-dispersed. An individual tree bears both male cones (as small as a little finger) or female ones (as large as a fist), but these develop at different times, facilitating cross pollination between trees.

A genus of thirteen species, *Agathis* is represented in Brunei by two species, *Agathis borneensis* and *Agathis dammara;* the latter occurs as isolated, though sometimes sizeable, populations on steep mountain terrain. Most other species are restricted to the Queensland and New Caledonia region. Before the species were properly distinguished by recent studies, both which occur in Brunei were referred to by the name *Agathis dammara,* in reference to the bark-resin *(damar* in Malay) copiously produced by these trees. Though no longer gathered for use in Brunei today, this resin has been used for fuelling torches. Still, *Agathis* trees are best known for their high-quality softwood, known in trade circles as *damar minyak;* this very value has encouraged the decimation of the most extensive stands in southern Borneo. Forestry in South East Asia now recognises the value of, and the threats to, *Agathis* trees and

Plate 8

there are programmes aimed at productive plantations of these trees and enrichment, by planting, of natural stands.

Platycerium ridleyi

A swamp staghorn fern Plate 9

Ferns of the genus *Platycerium* are probably the most regal among epiphytes. The different species, sometimes called *tanduk rusa* (deer antlers) in Malay and *lukud* in Dusun, are mostly large ferns perched on tree branches, developing sequentially bifurcating green fronds that resemble stag horns. These have become, understandably, much valued in the horticultural business and pressure from overly zealous and illegal collectors may easily destroy whole populations of the rarer species.

In Brunei's peat swamp forests *Platycerium ridleyi,* distinctive by its strongly ribbed basal fronds that wrap themselves like a cabbage around tree branches, and the stiff bifurcating "staghorn" fronds, can be frequently encountered as isolated small populations. Ferns do not bear seeds, but instead produce large quantities of minute, powderlike spore that germinate to produce tiny male or female forms vastly different in appearance from the larger green individuals we see, and which are usually inconspicuous. The mating between such forms produces an embryo that develops into the usual spore-producing plants. Although all structurally complex plants have this "alternation of generations", the ferns differ from flowering or seed plants in that the generations are physically separated, whereas both generations exist on the same plant body in seed plants. In *Platycerium ridleyi* thousands, perhaps even millions, of tiny spore-bearing sacs are borne on the lower side of specialised cuplike lobes from the base of the green fronds. As the spores mature, these sacs ripen and split, releasing to the breeze masses of spores on their way to a new life elsewhere.

Plate 9

Coelogyne monilirachis

The mountain bead orchid Plate 10

One of the most distinctive orchids of high hills and montane forest, *Coelogyne monilirachis* grows on tree stems and branches in moist, shady places. It is endemic to Borneo but is one of a large genus of some 150 species distributed from the Himalayas and southeast China to the Pacific islands.

Coelogyne monilirachis bears only a single attractive, wavy-edged leaf on each shoot, which at its base produces a similar new shoot. This habit gives rise to a series of such shoots and is termed a sympodial growth habit. Each individual shoot produces only one inflorescence at its tip, and this inflorescence bears a short row of maroon-brown bead-like segments (which gives rise to the name *monilirachis,* meaning "a stalk of beads") each of which produces a flower following a sequential order from the base upwards. After several flowers have been produced in this way, the slender inflorescence bears the current translucent pink blooms near its tip, and the whole inflorescence is arched over by the weight of this, nodding gently with the mountain breeze.

Many orchids grow in open places or on trees and have wind-dispersed seeds produced in large quantities. *Coelogyne monilirachis* is no exception, and its seed capsules split on ripening to release masses of tiny winged seeds which float away on air drifts to a new life elsewhere. Not all seeds will land on suitably moist surfaces on the bark of trees or in damp mossy patches, and the great amounts of seed produced help ensure that at least some will survive.

Several species of this important group of orchids are commonly cultivated in gardens in the region, including *Coelogyne asperata, Coelogyne mayeriana* and *Coelogyne pandurata.* The last named, found in Borneo, Sumatra and Peninsular Malaysia, is the "black orchid" popular in Europe, so called for the almost black markings on the largest (or "lip") petal of the flower.

Plate 10

Rhizophora mucronata

A mangrove tree

Plate 11

In Southeast Asia two species of the mangrove genus *Rhizophora* (or *bakau*) are common, and can even be found growing together. These two species, *Rhizophora apiculata* and *Rhizophora mucronata,* grow as straight, prop-rooted trees commonly forming pure stands in mangrove fringing estuaries and developing on tidal islands. Dense stands of *Rhizophora* invoke the comparison with a phalanx of soldiers as they stand stiffly together, even in the soft mangrove mud, seemingly to resist intrusion by other trees.

Rhizophora trees are even equipped to prepare the next generation of seedlings for a good start in life. Unlike many other plants, the fruits do not fall away as they ripen, instead the single seed within each fruit begins to germinate while it remains attached to the mother tree. Sustenance is channelled from the mother to the offspring as the seed germinates, producing a stout, green seedling stem (called a hypocotyl) that hangs from the mother tree, resembling a dangling long bean. This phenomenon is termed vivipary, comparing the process to birth in animals where the foetus derives sustenance from the mother before becoming physically separate.

In Brunei and neighbouring countries, *Rhizophora mucronata* is called *bakau kurap* ("warty *bakau*"), probably in reference to its warty seedling hypocotyls, and *Rhizophora apiculata,* with smooth hypocotyls, is called *bakau minyak* ("oily *bakau*").

Bakau poles are taken for scaffolding and piling in building construction, and remain a preferred source of wood in charcoal-making. More importantly, *bakau* stands stabilize river banks and mud, and contribute to coastline protection and natural land reclamation. They are important members of the mangroves which provide the breeding grounds for commercial fisheries, and wildlife such as the proboscis and leaf monkeys feed on their shoots.

Plate 11

Plate 12

Hoya coronaria

Milkweed of *kerangas* forests Plate 12

Hoya vines are well-known members of the milkweed family which are popular cultivated plants in Europe and America. There is a large variety of these vines in the lowland and hill forests of South East Asia, all of which, as their popular name suggests, produce a milky latex. *Hoya coronaria* occurs in dry, exposed habitats, widely distributed in the region, and in Brunei is frequently found in the white-sand *kerangas* forests near the coast, where their thick, rigid leaves seem well adapted against excessive water loss in a dry environment.

The beauty of *Hoya* vines lies in the glossy green leaves which occur in pairs and the attractive, rigid, waxy blooms which represent one of the most complicated flower structures known in the Plant Kingdom. Each individual flower has a specialised structure resembling a crown appropriately called the corona, in addition to the five fleshy petals which are joined at their margins to form a star-like corolla. The corona is a highly ornate crest formed by five fleshy lobes at the top of the flower, which shields the anthers and stigma of the flower. *Hoya* biology is equally interesting. Each inflorescence consists of a main stalk which bears at its tip a whorl of flowers that form a cluster called an umbel. These blooms last several days after opening and then fall off, after which a new umbel of flowers is produced by the same inflorescence stalk; each inflorescence stalk repeatedly produces flowering umbels and in its lifetime can give up to hundreds of blooms. The flowers of *Hoya coronaria* produce little scent during the day but are especially fragrant at night. This, and the pale colour of the flower which makes for better visibility at night, suggest that the flowers may be pollinated by nocturnal insects. The flower provides a reward to the pollinating insect in the form of copious quantities of nectar exuded from special glands in the flower.

Hoya vines resort to a special metabolism, called crassulacean acid metabolism (CAM) because it was first studied for succulents of the family Crassulaceae. Through CAM, the leaf stomata (pores) are open at night—when it is cooler and moister and excessive water loss through stomata is reduced—to admit carbon dioxide which is then converted and stored as malic acid. In daytime, the carbon dioxide is then released and used in photosynthesis, removing the need for wide-open stomata to facilitate gas exchange for normal photosynthesis. This is additionally efficient as during the night the vine can use up the respiratory carbon dioxide it and other plants produce. Orchids and ferns, many of which are epiphytes that live in dry conditions and need to conserve water, have adopted this kind of metabolism.

Life Forms

Trees are undoubtedly the main entity within the rain forest. Within the framework of trees, an explosion of other life forms occurs—there are lianas, palms, shrubs, herbs and bamboos, as well as a bewildering variety of epiphytes, creepers, mosses and liverworts. Some have developed specialised modes of survival, as stranglers or simply drawing their nutrition from dead organic matter. Others, including both trees and ferns, form specialised relationships with ants. Pitcher plants abound in some habitats, their expanded leaf tips forming curious pitchers which are death-traps for many insects and home to some animals. The variety of life seems infinite, spurred on by complex physical and biological interactions. Plants and animals compete among themselves and with one another, as they adapt to their physical environment. With increasing specialisation a multitude of life forms comes into being.

Competition and specialisation are not characteristics merely of human societies but in fact pervade every sphere of life. The rain forest, encompassing all its plant, animal and physical components, is a constant battle ground for survival. Green plants require sunlight, carbon dioxide, minerals and water as essential resources for growth. There are different amounts and combinations of each resource in the forest, from one spot to the next, and from the forest floor to the canopy. When a new genetic variant or form is produced, it may adapt to a set of resource conditions better than an existing form, and may be "selected" to survive in place of the original form, or together with the existing form at one locality, with each using different niches. Hence in a single area of the rain forest one can often find a number of closely related species of a genus, such as *Shorea* (Dipterocarpaceae) or *Syzygium* (Myrtaceae, the rose-apple family), many of which occur at low densities or as small dispersed populations.

In the lushness of tropical vegetation, plants not only have to compete for these resources and space but have to interact with animals. Effective reproduction and dispersal involves attracting the right pollinators and dispersers, and having pollens, seeds and perhaps even rewards packaged and presented effectively. At the same time plants are a source of food and those with features that can help defend themselves from being completely ravaged by the huge numbers of herbivores in the forest are able to survive better.

This is the foundation of the variety of form and behaviour among the plants and animals of the rain forest. Some plants develop fibrous, thick leaves that herbivores find difficulty in chewing through, or which last longer, say in a mountain habitat, where the lower temperatures and scarcity of nutrients mean that growth is slower and tissues are more "expensive" in terms of the plant's resources. Insects may then concentrate on eating the tender young shoots of these plants. Plant species producing a large flush of new leaves simultaneously, instead of gradually and continuously, may be able to satisfy their predators so that enough is left over. Plants also use chemical defence mechanisms as part of their armoury. The experienced forest traveller will see that a red exudate is produced by trees of the nutmeg family (which in Borneo are called *pendarahan,* or "trees producing blood"), trees of the mangosteen family may give a yellow latex on wounding, and others may produce a white latex—obviously a manifestation of different chemical properties. Thus the white latex of some species of the rubber-tree family, which coagulates into a glue-like substance on exposure to air, may simply deter damage by many insects by gumming up their mouthparts. As special chemical characteristics are selected in some groups of plants—which may be simply distasteful, or lethal to insects—and so serve as a kind of defence, herbivorous insects that develop some means of tolerance or immunity, such as by accumulating and excreting, or metabolically converting the chemicals (and so detoxifying otherwise potent poisons), will be able to feed on such plant material.

All this leads to greater competition between species which ensures that "the fittest shall survive" within a particular set of environmental conditions. This does not happen overnight. For amongst populations that are generally vulnerable to a new threat, or confronted with a new set of resources, only some individuals may, through the possession of slightly different characteristics, survive and persist. As these survivors reproduce, the very traits that have adapted them become part of the new population. This natural selection produces new, different forms, over millions of years. The process of change and adaptation in which two kinds of organisms, such as a plant and an insect herbivore, each influence the other, has been termed coevolution.

This can occur through other mechanisms as well. Plants have rewards, such as nectar, that attract potential pollinators but the specialised structures of some flowers may admit only certain types of pollinators which are perhaps favoured. The long-tubed, fragrant flowers of some *Ixora* trees may be amenable to nocturnal moths with long tongues, and the shallow, dish-shaped flowers of the *mempisang* family (Annonaceae) are probably well suited to the clumsy, scrambling movements of such pollinators as beetles and ants. Even mammals are not spared the lures of plants. The well known *petai* (*Parkia*) trees, which produce long pods of seeds considered a local delicacy, hold their tiny flowers on long-stalked, club-shaped inflorescences well clear of the crown foliage. As the flowers open at night a rancid scent is produced and visiting bats, as they feed on nectar and flower parts, have their breasts dusted with pollen. Banana (*Musa*), *durian* (*Durio*) and the mangrove *perepat* (*Sonneratia*) trees of Brunei are bat-pollinated, and may well be largely visited by bats from the cave complexes of Mulu and Niah in neighbouring Sarawak. Cantor's Roundleaf Horseshoe Bat (*Hipposideros galeritus*), a normally cave-roosting bat, is sometimes found hanging on palm fronds in the Sungai Liang Forest Recreation Park in Brunei. In Brunei, palm civets and hornbills are commonly known to be fruit-eating, and must be among the most important seed dispersers of forest plants. Traditional fruit gardens where trees bear abundant fruits on unobstructed bunches and stalks on the branches and trunks, such as the *rambai* (*Baccaurea motleyana*), *tampoi* (*Baccaurea macrocarpa*), *langsat* (*Lansium domesticum*), *tibadak* (*Artocarpus integer*), or on the edge of the crown in conspicuous bunches (the *pinanasan, Dacryodes rostrata; kembayau, Canarium odontophyllum;* and *rambutan, Nephelium lappaceum*) and singly (the *terap, Artocarpus odoratissimus*) are frequently defended from bats by erecting fine-meshed nets. Most forms of coevolution between plants and animals are of a loose nature, such as between a group of related plant species and a guild of insect herbivores, although there are examples of highly specific relationships, such as between particular fig and wasp species.

Trees and palms

The very structure of the forest itself creates a whole spectrum of environments which are exploited by a range of plants in different ways. Trees, with the capacity to produce woody, erect trunks that support a crown whose main function is to capture light, are the dominant life form of the rain forest. There are not just many species of trees, but also many *forms.* Different tree species develop to different levels in the framework of the forest, forming a variety of crown shapes. Even the same tree may develop an adult crown substantially different from that of its younger stages when the crown is progressively higher up from the ground as the tree grows taller. The deep, conical crowns of young trees of *kapur* (*Dryobalanops*) and *tulong* (*Agathis*) transform into wide, gigantic cauliflower-like structures or rather flat-topped crowns, respectively, when the trees finally break through the general canopy of the forest and their crowns are fully emergent. Each species, governed by its set of genes, develops according to a pre-set plan and even when the plan is upset (as when a growing crown is damaged by a falling neighbour) the growth responses to repair damage may also be pre-determined. In the 1970s botanists discovered that the thousands of different tropical trees can be categorized as following less than thirty different basic growth plans during their development.

A tree, by definition, need not be a huge structure. There can also be small trees only a metre or so in height. The main stem of a tree is invariably woody and erect, making them independent of support, and the foliage is largely carried some distance above ground, usually as crowns, by just one or only very few trunks. The vast majority of trees in the tropical forest are dicotyledons: flowering plants with two cotyledons, or seed leaves, in the embryo, and net-like leaf veins. A few are monocotyledons (flowering plants with only one cotyledon in the embryo), which frequently have leaves with nearly parallel rather than net-like veins, such as *Dracaena;* or gymnosperms (non-flowering plants producing seeds not enclosed within special fruit structures), such as the *tulong* (*Agathis*) or *jati laut* (*Podocarpus*). Palms and screw pines which are trunked are also considered trees by some, and referred to by others as arborescent (i.e., tree-like) monocots. This reminds us that, unlike these monocot plants, most forest trees are botanically dicots, which can substantially increase their trunk girth after growing tall.

Palms are certainly trees when they reach thirty metres or taller, as they can in the *nibong* (*Oncosperma tigillarium*), the *nibong sembaranai* (*Oncosperma horridum*) or the gigantic fan palms *bindang* (*Borassodendron borneense*) and *serdang* (*Pholidocarpus maiadum*). *Nibong* is common in brackish-water habitats, behind mangroves and in coastal mixed dipterocarp forests. *Nibong sembaranai* is frequently found inland even up to hilly country, whereas *bindang* and *serdang* palms in Brunei are more localised in 1owland alluvial or riverine sites. One should not forget the ubiquitous pan-tropical coconut palm. Resembling these arborescent palms in having a trunk with a large rosette of huge fronds at the top are the tree ferns, most abundant in hill and lower montane forest.

On the other hand, palms are clearly not trees when they exist as short-stemmed individuals or clumps bearing large, feathery or fanshaped leaves. The richness of the Brunei forest in palms provides many such examples. Among the most well known are the *tunjang pipit* (*Eugeissona minor*), *som som* (*Salacca affinis*) and *silad* (*Licuala*). The *tunjang pipit* begins life on the forest floor as a small feathery-leafed palm but begins to develop conspicuous stilt roots which, as the older, lower portions of the stem decays, appears to prop up the entire palm from the ground. If the shoot grows towards one side, more prop roots are developed there to support the crown of leaves, and sometimes side shoots begin to grow, similarly propped up. A mature *tunjang pipit,* where the original stem portions have decayed and disappeared, would have gradually "walked" away from the spot where it originally established. The *som som* (or *tuncum* in Dusun, and *redan* in Iban) is known for its astringent fruits which resemble those of the *salak* palm (*Salacca zalacca*). A clump of *som som* is but a short inconspicuous stem with huge leaves to three or four metres long. Each leaf has a strong, stiff midrib bearing on each side broad, somewhat S-shaped segments borne in groups and at different planes. The beauty of this palm from a distance disguises the threat posed by rows of sharp, though still ornate, spines at the bases of the leaf stalks. *Silad* palms (or *gernis* in Iban) are related to the New Hebrides fan palm (*Licuala grandis*) and the most commonly encountered species have leaves divided into wedge-shaped segments forming together a somewhat circular outline to the leaf, reminiscent of the spokes of a cycle wheel. These palms are stemless or short-stemmed ground dwellers of the forest, and the largest species have big leaves held on stalks several metres long.

Palms can be huge trunked plants, smaller understorey stemmed and stemless individuals, or even merely diminutive forms hardly a metre high, as some *Pinanga* and *Areca* species, often called *pinang* in Borneo. Another group, consisting of an entire subfamily of the palms, are the rattans, which are typically climbing palms strongly armed with spines on the leaf sheaths that clothe the stem. Generally called *rotan* in Brunei, Indonesia and Malaysia, these plants are frequently encountered in the forest with the prostrate older portion of stem (where the spiny sheaths have decayed and fallen off) crawling about on the ground and among trees, and the younger portion (protected by the tightly embracing spiny sheaths) suspended from high on the crown of a tree to which it has attached itself. Such a rattan must have lived a long time, for a young plant (or the young shoot of a fallen stem) can only attach itself to the foliage of a young tree by the hooks on the midribs of its leaves, or by specialised spiny whip-like structures, and so climb onto taller trees or ascend upwards as the young tree itself grows. Many a time the forest trekker has his hat or clothes caught by the spiny whips of a rattan; these climbing organs are generally called *unak* in Brunei and can be differentiated botanically as "cirri" if they develop at leaf tips, or as "flagella" if they grow out from the side of the stem sheath. It seems a great deal of trouble to designate these nasty structures with special names but groups of rattans can be identified from whether they produce cirri or flagella. Some rattans are non-climbing and stemless, and therefore do not produce these whips, and instead exist as rosettes of leaves. The gigantic rosette of *Daemonorops ingens,* as high as four metres and with huge pinnate (feather-shaped) leaves, is such a rattan and is as well-armed with spines on the leaf stalks as any other.

Of more than a hundred palm species in Brunei, several have interesting and appropriate local names. For example, *rotan air,* meaning "the water rattan" refers to *Daemonorops fissa* which is abundant in freshwater swamps and along lower streches of rivers in Brunei. *Rotan dahanan* ("the rattan with branches") refers to many species of *Korthalsia* canes that produce aerial branches in the forest canopy, a behaviour that is characteristic only in this genus. Likewise the Kedayan name for *Plectocomia mulleri* is *rotan menangis* ("the weeping rattan"), a massive rattan common in Brunei at the margins of peat swamps in the lowlands, aptly used for a species with slender hanging leaflets that drain off rainwater in streams from their tips.

Woody climbers

In Brunei, the lianas, or woody climbers, are a great deal more abundant in lowland mixed dipterocarp forest and in lower montane forest than in the other forest formations. On the ground, the lianas are mostly encountered as thick, rope-like stems twisting and winding their way into the forest canopy, and occasionally as sinuous portions which have fallen to the forest floor. Lianas hold most of their foliage in the crowns of trees, where light is more abundant. They are in many ways sun-loving plants, having to reach for better light conditions in the forest canopy and many begin to establish (by seeds, or through sprouting from fallen stems) in the bright forest gaps created by a tree fall or a tree having died standing. Liana growth can be so prolific that such openings in the forest are smothered; and, as young trees establish and grow, the canopy of tree crowns intermingled with the lianas is gradually raised. In this way many lianas grow, with trees, high into the forest canopy. The thick, old lianas seen in the forest can therefore be as old as the largest trees. The role of liana growth in large forest gaps may be an important one ecologically, as it helps provide a partly sheltered environment in which seedlings and young trees intolerant of full light can establish themselves.

Liana seeds do not always land in light gaps in the forest, nor must they necessarily depend on bright conditions to germinate. In the deep shade of the forest canopy, liana seedlings are often found developing and some, like young saplings of several species of tree, may persist in a stunted form resembling treelets until more light breaks through the canopy to stimulate further growth. Thick liana stems can drape the tallest tree trunks and frequently hang across from one tree to another, and may hang down to the forest floor. These connections provided by lianas between trees serve as bridges for small mammals such as rodents in the forest, and can be so strong that the falling of one tree can pull down other trees or tear off huge limbs.

Lianas get onto trees and branches through a variety of devices. Some, like the gambir plant *Uncaria,* use hooks on the branches while others, like the *Tetrastigma* grapevines, employ tendrils. Still others, such as a number of climbers in the *mempisang* family (Annonaceae), have the main shoot tip curl around a support, while a side branch develops to continue height growth; it too eventually grasps a support and is similarly replaced by developing its own side shoot, so that a series of shoots brings the liana plant progressively higher. Certain other lianas have their young, still-tender stems twine around small trees, sometimes aided by very rough stem surfaces for maximum friction, as in the Sandpaper Vine *Tetracera.* Whatever the method of getting up to the canopy, the stem of a liana is a wonder of nature, a flexuous conduit of tissues transmitting water and nutrients from the roots to the highest shoots, and distributing the products of photosynthesis from the leaves.

Lianas (*akar* in Malay) are also familiar through their usefulness to people in Brunei. Short lengths of *akar engkelait* (*Uncaria*) and *akar mempelas* (*Tetracera*) stems yield potable water often taken by forest travellers, and good ropes are fabricated from the stem fibres of the *tengang* (*Gnetum*) climber. In the *tamu* or open markets of the towns, lengths of liana stems are sold for their medicinal or magical properties. For example, stems of *Securidaca inappendiculata* (*langir* in the Brunei language, *endaki* in Dusun and *pakar* in Iban) are sold for their saponiferous quality when mashed up and immersed in water used for washing in the treatment of skin ailments and after childbirth, and as a shampoo. The slender tubercled stems of *Tinospora crispa* (locally called *ratnawali,* a name derived from the Sanskrit meaning "a string of jewels") are also well known in their use for treating a variety of ailments including malaria, bacterial abcesses and high blood pressure. Some lianas of the family Aristolochiaceae are known among naturalists to be the preferred food of caterpillars of the spectacular Birdwing butterflies, now being threatened by overzealous and commercially motivated collectors in many places in Borneo, and the Far East.

Other climbing plants

Somewhat akin to the lianas are the slender climbers and vining bamboos. Some ferns have developed climbing stems, exemplified by the common *lamiding* fern (*Stenochlaena palustris*) whose young fronds are savoured as a vegetable in Brunei. Another group of ferns, in the genus *Lygodium,* has creeping subterranean stems and the climbing, twining parts are in fact the midribs of the leaves, giving a growth habit unique among ferns! The herbaceous *rotan buntak* (*Flagellaria*), with pronounced leaf-tip tendrils, is not a rattan but a climbing monocot in a family of its own. Vining bamboos, though unheard of in many places around the world because most bamboos are erect clumps or groves, are well known in Brunei as *buloh badan.* Northern and northwestern Borneo, including Brunei, is a centre of diversity of *Dinochloa* vining bamboos, which develop slender stems and branches that actually twine around the stems and trunks of other plants.

In many ways, vining bamboos contrast with erect bamboos with which many people are familiar. Erect bamboos in the tropics thrive in open, cleared sites or along exposed riverbanks and can invade a particular site by the sheer number of clumps growing densely together. In such places the dense bamboo growth shades out many plants and the accumulation of its slowly decaying, siliceous fallen leaves on the ground prevents other seedlings from reaching mineral soil easily. On the other hand, vining bamboos send out slender shoots in the forest understorey and twine their way up tree trunks, infiltrating a stand of trees from within. The different growth habits and invasive behaviours of erect and vining bamboos have been likened to phalanx and guerilla strategies in warfare, respectively, because one form thrives by keeping out other plants through multiplying as dense stands and the other by infiltrating an existing stand of trees. In large gaps and disturbed forests, vining bamboos grow quickly to form a layer over the other plants; their effects on tree regeneration in such places is not yet fully understood.

As with bamboo, the existence of both erect and climbing habits is also seen within the screw pine (*pandan*) family. Members of this family are easily recognised by their leaves with prickly edges and which are arranged in three rows that spiral up the stem. The principal genus of the family is *Pandanus,* including the stemmed, tree-like and prop-rooted forms called *pandan* or *gadur* in Brunei, and the large stemless forms called *mengkuang.* Another genus, *Freycinetia,* includes climbing species often festooning old tree trunks in moist places.

Shrubs, rheophytes and ant-plants

Shrubs are woody plants which do not develop main stems that are recognisably trunks supporting most of the foliage. Instead, they produce many stems that are usually small where the foliage is well shared among all the stems. Many shrubs can therefore attain a bushy appearance, like stunted tree crowns devoid of a trunk and arising directly from the ground, although some can be small with only a few leaves. Shrubs are comparatively rare in lowland mixed dipterocarp forest but seem to be common where abundant light reaches the ground, such as in *kerangas*, stream and river banks and in upper montane forest. In *kerangas* sites with sparse vegetation, the grey-green bushes of *Styphelia malayana* (*terindak ati*) and the curious fig plant with triangular leaves, *Ficus deltoidea,* are common. On the highest mountain ridges, shrubby life forms abound, and many belong to the heather family Ericaceae, including rhododendrons and species of the blueberry genus (*Vaccinium*). These are so conspicuous that upper montane forests are often also known as Montane Ericaceous Forests. Along the pristine, rocky streams and rivers of the lowlands, some shrub species may be conspicuous. Both *Osmoxylon borneense,* a remote relative of the famed ginseng plant, and *Myrmeconauclea strigosa,* of the coffee family, are frequently found along fast-flowing streams and rivers, although the latter also grows in bouldery narrow streams. Shrubs which are confined to rivers and streams are part of a group of plants which are similarly adapted, called rheophytes. Rheophytism is an adaptation to life in or along rivers and streams and does not define a special growth form. The huge dipterocarp tree *Dipterocarpus oblongifolius* (*neram* in Malay and *ensurai* in Iban) is only found along the middle stretches of large rivers and is a rheophyte, as are the tiny plants which live in rock crevices on river banks or are submerged in rivers. In general, rheophytes growing on the banks of rivers or among rocks in the river develop a strong root system which serves as an effective anchor to prevent being washed away. They can tolerate occasional submergence of some or all of its parts, and are frequently characterised by narrow or finely dissected leaves which presumably offer less resistance to river currents in comparison with broad leaves. The *ensurai* trees that tower over the river, arching in places to form a canopy over the water, may look like many other forest trees, but young *ensurai* plants bear remarkably narrow leaves in comparison to the adults. At times, even the lowest branches of these huge trees, suspended just above the water, will have unusually narrow leaves.

The rheophytic shrub *Myrmeconauclea strigosa* is interesting in another way. Called *momban* by the Iban people, this plant develops swollen branch segments which, on examination, have tiny holes made by the black ants which reside within. Plants which develop special associations with ants are called myrmecophytes or, simply, "ant-plants". Ant association, like rheophytism, is a special adaptation rather than a special growth form. The ant-plant is defended by the ants when the plant is disturbed, for example by the presence of a hungry caterpillar. In return, the ants are provided with a place to nest or breed. The plant may merely offer materials and a suitable shelter for ants, as in the rattan *Daemonorops formicaria,* where the ants build long surface runways of interwoven fibres from the fine fluffy material gathered from the rattan's young shoot, which are well protected from the probing tongues of mammals by the whorls of spines on the leaf sheaths. However, ant-plants may be more specialised, forming enlarged tubers or rhizomes riddled with interconnected tunnels housing the ants, such as in *Hydnophytum formicaria* and *Myrmecodia tuberosa*, which are common in *kerangas* and peat swamp forests. Some ant plants, such as the *sedaman* (*Macaranga*) trees, may offer additional incentives. Several *sedaman* species produce food bodies on the young leaves or the stipules (leathery small structures at the stem nodes) which the ants take. The ants live inside the hollow stem segments, where they derive sugary substances from scale insects that they husband and which suck the tissues of the stem.

The plant, too, may derive more from its association with ants than just protection from herbivores. In experiments using radioactive tracers, inorganic nutrients from the detritus gathered by ants have been shown to be absorbed by the plant. The *Dischidia* climbers, of the milkweed family, develop special enlarged, "pitcher" leaves where the ants nest and into which roots grow, produced from the stem to withdraw nutrients from the decaying material within. Sometimes "ant gardens" develop, when seeds of certain plants are carried by ants into a special nesting site, such as the pitcher leaves of a *Dischidia* plant, and begin germinating in the rich organic material gathered there.

Perhaps some of the most dramatic examples of ant-plants are the rattans *Korthalsia hispida* and *Korthalsia robusta.* The rattans themselves are a curious sight, with the stem tightly clothed in the spiny sheaths, each of which has an inflated upper portion cupped against the stem. The ants, which live in these chambers, become highly agitated if the rattan is struck, and a current of rattling sounds travels upwards along the cane as the ants bang their abdomens against the chamber wall.

Herbaceous plants

In contrast to the trees, shrubs, lianas, palms and bamboo are the herbaceous plants of the forest floor, which are soft-tissued throughout. Ground herbs may be small leafy plants such as begonias and other species from many different families, including the coffee family Rubiaceae and the taro or *keladi* family Araceae. Although the variety of herbs is great, there is not an even cover of herbs on the forest floor, with dense patches of herbs occurring especially in moist spots or small forest gaps. Particular herb species may be highly localised in their distribution, as are many species of the African Violet family (Gesneriaceae), many of which are small rosette herbs with violet blooms. Some herbs are aquatic, including the rare bottom-rooted rosettes of *Hydrostemma* (water lily family) and *Cryptocoryne* (*keladi* family) which reside in exposed pools and slow-moving water channels of freshwater swamps. The leaves of these aquatics float on the water surface or are submerged, held by long leaf stalks from a short stem rooted in the bottom. The largest ground herbs, sometimes referred to as "giant herbs", may be two to three metres tall and include the forest gingers and bananas which are most frequently encountered beside rivers, in large forest gaps and fringes. While some gingers, such as *Boesenbergia* or *Globba,* may be less than half a metre high, banana plants are as a rule larger, with false trunks which are actually built up of overlapping leaf stalks. A number of ground ferns resemble the soft-bodied herbs but many have hard, brittle creeping stems and leaf stalks.

Saprophytes, fungi and other lower plants

Some ground herbs are devoid of any green chlorophyll pigments, and are therefore unable to carry out photosynthesis, and are often found in heavily shaded, moist spots. These herbs are often tiny, with their leaves reduced to minute scale-like structures or even absent. They obtain nutrients through their roots forming an association with fine fungal threads that are able to break down the lignin and cellulose material of decaying plant matter. Plants that utilise such a relationship are called mycotrophic saprophytes (or "plants that draw nutrition from dead material through the feeding of fungi"). Very often a mycotrophic saprophyte is almost entirely roots and a stalk of flowers. The fine purple-flowered tufts of *Sciaphila,* the slender pale-flowered individuals of *Burmannia* and the wiry, solitary *Aphyllorchis* (a leafless orchid) are such saprophytes.

Fungi, however, are the true saprophytes, for they possess enzymes capable of breaking down lignin and cellulose. Although fungi are distinctive and remembered for their special spore-bearing structures called mushrooms and toadstools, most of these plants exist as a mycelium—a network of fine fungal strands often microscopic in dimension. On the forest floor, fungal mycelia are the main agents of decomposition, a necessary process in releasing the nutrients held in organic matter so they can be available again. As mentioned, fungal mycelia also form special associations with the roots of plants, called mycorrhizae, where they form a sheath over the finest root tips, or penetrate the superficial root tissues. These fungal root sheaths assist nutrient transfer from the fungus to the root, whereas penetrative mycelia within roots form nutritious globules or thickenings which gradually become assimilated by the root cells. Dipterocarps have been found to be associated with fungi forming such sheaths around their root tips indicating that these gigantic forest trees derive nutritional benefit from fungi. Orchids, too, depend on fungal associations to survive. Their seeds are tiny and do not store much reserves, and most require a fungal association to channel enough nutrition to the seedling before it develops enough green tissue for photosynthesis. The associated fungus enters the orchid seed even before germination and penetrates the young embryo. There is no doubt, however, that the spore-bearing structures called mushrooms and toadstools, sometimes brightly coloured, are the most attractive parts of a fungus. Often, dull-coloured spore-bearing bodies in the shape of thick brackets protruding horizontally from rotting logs or dead standing trunks, produced by *Ganoderma* fungi, are very noticeable too. Some fungi produce mushrooms that are bioluminescent, naturally producing an eerie glow in the night, but sometimes leaves and twigs, infested with mycelia engaged in their decomposition, also emit a faint glow.

As a group, the fungi are not as developed as ferns and flowering plants and their plant body and tissues are comparatively simple in organisation. Often grouped together with the fungi as lower plants, are algae, which not only form greenish or brown strands in the rivers but are also visible as an orange fluff on some rocks and the tree trunks. Lichens, formed from the interweaving of algal and fungal tissues and behaving as single (compound) organisms, are often visible on many tree trunks as patches and hoops of grey, green, red, black and even blue. The associations are often quite species-specific. Lichens not only form colourful patches but can also take on special forms, such as the Old Man's Beard lichen (*Usnea*), which forms tufts of grey-green strands on the branches of trees in montane forests. Lichens can also assume crustlike, coral-like and leaf-like forms, or furrowed lobe-like structures. In a lichen the algal partner, which can photosynthesise and even fix atmospheric nitrogen, supplies the nutrition while the fungal partner is mainly responsible for the mass and shape and carries out sexual reproduction. Even more common and visible in montane forests, and also in damp places in lowland forests especially beside rivers, are the mosses and liverworts. These form sods on the ground and around trunk bases, and cushions and wefts on the trees. Thick beds of moss can build up in the highlands, in which many seedlings, ferns and other herbs become established. There the frequent cloud and mist cover keeps the moss and liverworts well-watered. Some mosses are common in the *kerangas* and peat swamp forests of the lowlands as well, where they also form soft moist beds at tree bases where small plants grow.

Epiphytes and parasites

Away from the forest floor, even on the highest tree branches, a totally different life form exists. Epiphytes, or plants wholly growing on others for support, are found among many different families and may be herbaceous or woody, and includes even the algae and lichens growing on leaves and tree trunks. Orchids and aroids (plants of the *keladi* family Araceae) provide many examples of herbaceous epiphytes. Though many epiphytic orchids are comparatively small, there are notably large ones. The orchid *Cymbidium finlaysonianum* is a large epiphyte developing a tuft of long narrow leaves. In contrast, the Tiger Orchid *Grammatophyllum speciosum,* develops into huge clumps on big trees, and counts among the largest orchids in the world. The aroid *Scindapsus borneensis* forms a large fan-shaped array of leaves against the tree trunk on which it grows, although it begins life as a ground-rooted creeper. As the aroid stem grows longer it bears larger leaves and finally the old basal part dies, leaving the mature upper portion attached to the tree trunk. The large fan of leaves is effective in collecting falling leaves, twigs and bark from the canopy higher up, so accumulating decomposing organic matter into which its roots grow. This litter-trapping growth habit, which helps supplement the nutrition, is also characteristic of large fern epiphytes such as the Bird Nest Ferns (*Asplenium nidus* and *Asplenium musifolia*) which form rosettes of broad, long leaves. Another epiphytic fern, *Drynaria quercifolia,* develops two kinds of leaves: litter-trapping nest leaves which are broad, oval, short-stalked and held steeply against the tree branch on which the fern is creeping, and normal leaves which are deeply lobed and long-stalked. Ants frequently nest in the debris collected by litter-trapping epiphytes.

Woody epiphytes include species of *Schefflera* (ginseng family), *Medinilla* (*kodok-kodok* family), *Fagraea* (*tembusu* family) and *Rhododendron* (heather family). In April and May the tubed or bell-shaped flowers of a number of rhododendrons can be picked up here and there on riverbanks in the lowlands, just below these epiphytes nested high on overhanging branches. Epiphytes depend on the "host" plant only for support and the majority do not appear to cause any ill effects on the host except in the way of causing branches to break due to their weight. Some fast-growing creepers, such as ferns, are said to smother branch tips to an extent they reduce the effective photosynthesis by the foliage there, causing branch death. This condition has been called epiphytosis.

More extreme are the truly parasitic epiphytes, which sink specialised absorbing tissues called haustoria into the host limb from where they withdraw nutrition and sap. These parasites can themselves bear green leaves or twigs so they can photosynthesize, but closer examination reveals points of contact and fusion between the tissues of host and parasite. Such are the hemi-parasites ("half parasites") of the genera *Dendropthoe, Macrosolen* and *Scurrula* in the specialised parasitic family Loranthaceae, and of *Viscum* in the mistletoe family Viscaceae.

The flowers of many hemi-parasites in the Loranthaceae are often tubular, with bright yellow or red colours, which are known to respectively attract bees and birds. These are thought to be the main pollinators of such parasitic plants. Birds certainly take the berries of these plants, and occasionally the germinating seeds of a parasite are found plastered in bird excrement on a tree branch. Parasitic plants are common but by no means limited to nutrient-poor habitats such as *kerangas*, peat swamps and montane forests. The leafy parasites contrast with the holo-parasite ("total parasite") *Rafflesia,* which is devoid of stems, leaves or true roots and is represented solely by a huge, gaudily speckled red flower and haustorial tissues. *Rafflesia pricei* (first recorded in Brunei in 1990) is, like other species of the genus, parasitic only on the stem base and roots of certain species of *Tetrastigma*, which are lianas belonging to the grapevine family Vitaceae. Related to the *Rafflesia* but with smaller flowers, and also found in certain spots in Brunei's forests, is another holo-parasite *Rhizanthes lowi*.

Litter trappers

Like epiphytes high in tree crowns which can trap falling litter to form a ready resource of organic matter, free-standing plants rooted in the forest floor can do the same. The leaf rosettes of young trunked palms such as the *serdang* (*Pholidocarpus maiadum*) and some *silad* (*Licuala*) effectively funnel litter falling from the canopy and accumulate a pile of decaying organic matter at the base of the palms. The fine roots of neighbouring trees grow into these litter piles and even ascend to the mass of litter carried in the young palm crowns. As the palm trunks grow taller the crowns and their trapped litter content are raised higher, and the ascending tree roots keep growing further up, zig-zagging their way between rotting leaf bases below the crown. Finally, large palms may be seen to carry several thickened zig-zag roots as high as one metre (in *Licuala*) or three metres (in *Pholidocarpus*) emerging from the ground and strung about the palm trunks, their distal ends nested in the crown. There is no direct evidence that such tree roots are adapted specially in nutrient-poor sites to exploit a rich organic resource carried by the litter-trapping palm crowns. However, as in Peninsular Malaysia where this has been observed, the phenomenon of the roots ascending palm trunks and nesting in their crowns is only common in Brunei in some moist, slightly peaty sites on sandy, nutrient-deficient soils. In other instances, free-standing treelets of *Trigonostemon* in the rubber-tree family form litter-trapping rosette crowns and send fine roots developed from their own stems into the accumulated organic matter. These curious treelets, however, are not especially noted in nutrient-poor sites.

The figs

Strangling figs (*Ficus*) provide one of the most unusual tree forms in the rain forest. A strangling fig begins life as an epiphyte when its seed is deposited by a bird in the crown of a tree, perhaps at a fork between two branches. There it grows into a woody shrub and sends down to the ground one or more roots which gradually thicken and branch, and fuse with one another upon contact. This interweaving of strangler roots around the tree increases in thickness and density, encasing the trunk tightly until the tree dies, its bark more or less constricted by the fig. Having strangled its host tree, the fig begins to take over, building up its own crown and hollow trunk-like casing. Some stranglers merely outlast their host trees by shading out the original crown, and do not constrict the host's bark. In some species aerial roots are sent down from the base of the fig's crown, developing on contact with the ground into large pillar roots that surround the trunk. These unusual trees are imposing and in most places in the Far East and the Pacific there have sprung up all sorts of legends and beliefs that such figs house spirits, which may sometimes be the soul of someone living. The spirit, or *antu,* according to Iban belief, can eat souls trapped by the aerial or pillar roots of a large fig, and a person whose soul is so taken will fall sick and die. Fig trees are generally known as *ara* in the region, and in Iban terminology strangling figs are called *ara pakau,* while large fig trees with columnar roots from the main branches are known as *ara ulur.* There is even an Iban term *ara nyadi* for the huge free-standing figs that develop spreading crowns in the forest, a comparatively rare sight compared to stranglers. The Dusun people refer to figs as *nunuk,* and also consider strangling figs as dwellings of spirits. As far away as Sicily fig trees are associated with spirits.

Fig trees are interesting for several more reasons. Firstly, the fig "fruit" is in fact not a single fruit, but is a hollow structure in which tiny flowers cover the entire inner wall. Botanically called a syconium, a fig has an aperture at its tip guarded by small scales, through which tiny wasps can enter or leave. The relationship between a fig-wasp and a particular species or several related species of fig is a highly specialised one. Many figs have three sorts of flowers: male, with anthers only; female, with an ovary and a long style; and gall, with an ovary adapted for nurturing a wasp's egg deposited through a sufficiently short style. Emerging male wasps from gall flowers are wingless and die shortly after mating with females; they do not leave the fig. Males of some wasp species inseminate unemerged females through openings which they bite in gall flowers containing females, and those of other species copulate with emerged females. Female wasps, however, are winged and after mating they leave the ripe fig, now no more rigid, through the aperture, making their way easily past the soft scales. As she leaves, the female wasp carries pollen deposited on it by the male flowers, and she brings this to a different, younger fig of the same species. In the process of struggling past the stiffer scales at the aperture of this young fig, a wing or antenna may be lost. When she is inside, the female wasp attempts to lay her eggs by inserting a slender ovipositor at her hind end through the styles of female and gall flowers. Egg-laying into gall flowers is successful because the styles are short enough, but with female flowers it is thwarted by the longer styles and in the process of trying the wasp only manages to pollinate the female flowers. Finally done with her attempts at laying eggs, the female wasp dies within the fig, leaving her larvae to develop within the successfully penetrated gall flowers, from which a new generation of wasps will emerge when the fig ripens. In some species, individual trees bear male and gall flowers together in a fig or only female flowers; this ensures cross-pollination between trees.

These pollinating wasps belong to a group called the Chalcidoidea and can only develop within the gall flowers of the fig; the fig flowers, on the other hand, can only be pollinated by these wasps. Studies have shown that each fig species has a specific wasp pollinator. The relationship would be quite smooth were it not for other Chalcid wasps that oviposit only into fig ovaries already occupied by eggs of the usual pollinating wasp, thus robbing it of gall ovaries it has stimulated to develop. In figs of the strangling *Ficus xylophylla* in Brunei, pollinating *Waterstoniella* wasps and parasitic *Grandiana* wasps have been identified.

Another fascinating point concerning the habit of fig species is where the figs develop on the tree. Many species bear the figs in the axils of leaves in the crown, where they are frequently taken by birds, but some produce them in clusters on short tubercles on the trunk, a phenomenon called cauliflory. In some species, the figs are borne predominantly at the base of the trunk, and this is termed basiflory. The most extreme situation occurs with the so-called "earth figs", where the figs are found on long slender runners from the base of the trunk; this is referred to as geocarpy. This diversity in flowering habit continues to puzzle botanists, although perhaps visible fig clusters on a bare stem are more attractive and accessible to certain mammals. Certainly what happens to geocarpic figs in the way of pollination (especially when the figs are somewhat buried) and dispersal requires to be better investigated. Basiflorous figs are called *entimau bandam* in Iban, while geogarpic figs are referred to as *entimau* in Iban and *lagenit* in Dusun.

Pitcher plants

Pitcher plants (*Nepenthes*) are undoubtedly one of the most interesting groups of rain forest plants, earlier mentioned as being common in nutrient-deficient environments. These climbing plants, called *somboi-somboi* in Brunei, can develop unusual forms of pitcher at their leaf tips but in general the ground pitchers are more squat in appearance and have two wing-like keels running down their sides. The aerial pitchers, borne higher up as the plant climbs, are usually more slender and devoid of the keels. Exceptions include the ultimately epiphytic pitcher plants such as the montane *Nepenthes veitchii,* left high above the ground following decay of their lower climbing stems on mossy tree trunks and branches, and not easily noticed. The lid of a pitcher exudes nectar that is attractive to insects which find the pleated smooth margin of the pitcher too slippery, and fall inside to their demise in a watery solution containing enzymes secreted by glands on the inner wall of the pitcher base. This enzymatic soup is able to break down proteins and the component nutrients are then absorbed by the pitcher. In spite of this, small frogs and some mosquito larvae have been known to blissfully survive in the pitchers; they are part of the so-called "infauna" of the pitchers.

Such is the life in the rain forest—varied, rich and complex. Much remains to be understood about the different life forms, some of which have fascinating aspects of biology. Yet, hundreds or thousands of species can coexist in the different forest communities, adapted to a range of physically and chemically defined environments, and their survival dependent on the processes of competition, adaptation and change. These are the wonders of the rain forest, unchallenged in diversity by any other ecosystem on earth.

Plate 13

Nepenthes lowii

Low's pitcher plant Plate 13

The *Nepenthes* family is one of three plant families in the world in which the plants develop pitchers that trap mostly insects. There are some 80 species of *Nepenthes,* distributed from Madagascar, the Seychelles, Sri Lanka, and Assam to northern Queensland and New Caledonia. Some 30 species alone are found in Borneo, of which about half occur in Brunei, where they are called *somboi somboi* (Malay), *tuyud* (Dusun) or *periok kera* (Iban).

Nepenthes are climbing shrubs with tough stems and pitchers at the hollowed-out tips of the leaf midrib tendrils. The leaf pitchers, undoubtedly the most interesting parts of the plant, were considered by Linnaeus, when describing the genus, to be such astonishing structures that they command full attention away from all other thoughts. He thus gave the name *Nepenthes* which in Greek means "to banish grief", in allusion to a drug in Homer's Odyssey which Helen cast into wine to free men from anger, grief and all ills.

Nepenthes pitchers are some of the most specialised structures developed in the plant kingdom. A lid over the top of a young pitcher prevents rain from washing out the contents and diluting enzymatic juices produced within. When it opens insects are attracted to the lid where there are usually distinctive colour patterns and nectar exudes from glands on its lower surface. Some insects may also be attracted to nectar glands at the base of the rim (or peristome) fringing the pitcher mouth. From these places many insects fall in and have great difficulty in getting up along the smooth upper part and rim of the pitcher. At the bottom of this trap numerous glands (as many as several thousand in one square centimetre!) secrete enzymes that break down the exoskeleton and proteins of the victims, which are further digested by bacteria into amino acids that are then absorbed through the pitchers walls. The glands in the pitcher also secrete benzoic acid which helps prevent rapid decay of the prey, and juices with acid values as low as pH 2 have been determined. Pitcher plants thus obtain their nitrogen supplement with these artful devices, and are thought to be able to adapt to nutrient-poor habitats because of this peculiar mode of nutrition.

Nepenthes lowii is perhaps the most exquisite of pitcher plants, first recorded by Hugh Low in 1851 on Mount Kinabalu in Sabah. Low's pitcher plant is only present on the higher mountains of Borneo, including the Retak and Pagon ridges in Brunei's Temburong district. Its gourdshaped pitchers, with an unusually slender "waist", are distinctive, but little is known of how they function or why the upper pitchers should have such a flared upper part and a slender waist.

Nepenthes bicalcarata

Pitcher traps of the peat swamps Plate 14

The leaf-tip pitchers of *Nepenthes* are not only fascinating as structures designed to attract and trap insect prey, but are also intriguing in the variety of form found from one species to another. The pitchers of *Nepenthes bicalcarata,* found in peat swamps in Borneo and in particular where the dipterocarp *Shorea albida* is abundant, differ from those of the highland *Nepenthes lowii* in the absence of a "waist" below the mouth, and in having two fang-like structures at the base of the lid.

Even on the same plant, for most species, the ground (basal) pitchers differ from the upper (aerial) pitchers. Ground pitchers are often more squat, and have two wing-like keels developed down their front; the leaf tendril passes straight down the front of the pitcher. In contrast, aerial pitchers are more slender, and bear only two ribs down their front; the leaf tendril makes a loop which often coils around a support and joins the pitchers from behind.

As clever as they may seem, the pitcher traps can be cheated. In Brunei, the Red Crab Spider (*Misumenops nepenthicola*) spins its web above the fluid within pitchers of the lowland *Nepenthes gracilis* and *Nepenthes rafflesiana,* intercepting insects (such as flies, ants and bees) as they fall inside for its own consumption, and dropping what it discards into the slimy grave below. The spider is also known to haul up and feed on mosquito larvae living in the pitcher fluid. Apparently, it can also escape predators by dropping into the enzymatic fluid itself, later to emerge

Plate 14

unharmed, although generally the silk threads it spins serves also for it to avoid falling in easily. Some other animals seem tolerant of the acidic digestive juices of the pitchers, and frogs, tadpoles, filter-feeding mosquito and fly larvae, and even small land crabs have been known to survive within the pitchers of various South East Asian species. These small animals residing within are collectively known as the "infauna" of the pitchers.

Nepenthes bicalcarata can also develop associations with certain ants. In some cases, an ant resides in the hollowed-out swelling on the tendril near where it joins the pitchers. In other instances, a Drummer Ant beats its abdomen against the lid of the pitchers, apparently a display of aggression, when the plant is disturbed; this ant also helps itself to nectar on the lid.

Pitcher plants produce male and female flowers on separate individuals, which ensures cross-pollination and thus outbreeding, another clever ploy. The leathery seed capsules on a female plant eventually split to release large numbers of light, wind-dispersed seeds.

Plate 15

Baccaurea racemosa

The Mata Kunau tree Plate 15

Mata kunau trees belong to the important genus *Baccaurea* which is frequently encountered in lowland dipterocarp forest. They are related to the well known *tampoi* (*Baccaurea macrocarpa*) and *rambai* (*Baccaurea motleyana*) fruits of South East Asia, both of which occur in Brunei. The *rambai* is widely cultivated for its fruits, the *tampoi* sometimes is also planted, and the *mata kunau* illustrated here (*Baccaurea racemosa*) is known only in its wild state.

These species share certain characteristics. All have the fruit bunches borne on the trunk and the branches behind the leaves, rather than among the leaves, a character that can only be found among some tropical trees. The fruit of these species bear seeds enclosed in a fleshy jacket, which is edible. The flesh of *tampoi* and *rambai* fruits remains a translucent pale to creamy colour, whereas that of the *mata kunau* eventually ripens blue. *Mata kunau* fruits ripen a bright red, and hang in short strings from the branches and trunk. The fruits are known to be eaten and dispersed by monkeys in the forest.

All these species are also dioecious, a condition where individual trees bear only male or female flowers, thus ensuring cross-pollination between trees. However, this condition adds greatly to the difficulty in conserving adequate populations of wild trees of species that exist scattered through the rain forest, because the conservation of any viable population must then include enough numbers of both sexes for adequate breeding.

Twenty five species of *Baccaurea* have been documented for Borneo, though there may be more as the inventory of species in Bornean forests progresses. More than one species are referred to as *tampoi, rambai* or *mata kunau* according to certain criteria, although those mentioned here are the best known. *Tampoi* trees are mostly larger, have red bark and big fruits, and *rambai* and *mata kunau* trees are mostly smaller, have brown or buff-coloured bark and smaller fruits. *Mata kunau* trees are frequently small and often have branches clearly expressed as a wavy series of shoots, each succeeding the previous one outwards. *Mata kunau* trees are called *engkuni* in Iban and *sekunau* in Dusun.

Melastoma beccarianum

A wayside shrub Plate 16

Common on the hot, dry degraded soils of wayside places and exposed sites in South East Asia are a handful of shrub species belonging to the genus *Melastoma,* of which perhaps the best known is the so-called Straits Rhododendron, *Melastoma malabathricum*. This name has come about because of the superficial resemblance of the large, showy blooms of *Melastoma* to some rhododendron flowers, even though the plants belong to different families.

Melastoma beccarianum, named for the great Italian naturalist Odoardo Beccari whose plant and animal collections, and writings in the 19th century have contributed greatly to science, has among the most attractive flowers in the genus. Known in Brunei as *kodok-kodok* and *kemunting* (a name referring to a different plant elsewhere in the region), the species flowers freely and the purple, large-petalled flowers are highly conspicuous. One of the most distinctive parts of the plant is the rounded basal portion of the flower, called the hypanthium, which is densely covered in red to crimson stiff bristles that each (inset, top left) ends in a tuft of spreading hairs. Also exposed in the open flower are the peculiar stamens (inset, bottom right, showing frontal and side views) in which each yellow anther is joined to a pink-yellow filament by a characteristically kinked connective portion.

Though species like *Melastoma beccarianum* have potential value as an ornamental plant, more so because it is adapted to harsh soils in open conditions and so portends ease of cultivation, in general plants like it fulfil an important ecological role. The herbs and shrubs in open, degraded areas form the pioneer community which covers the ground and encourages the development of other species which cannot establish directly on hot, dry sites. In time, the herb-shrub community grades into a secondary forest and, over a long time, primary forest plants begin to establish. Secondary forests are species-poor and their chances of being transformed into species-rich, tall forest depend on many factors, among which is the extent of degradation suffered by the site. In some places in the tropics, degraded sites and secondary forest are fast becoming larger in extent than lowland primary forest.

Plate 16

Arachnis calcarata

The epiphytic Scorpion Orchid Plate 17

Related to the popular scorpion (or spider) orchid *Arachnis flos-aeris,* the Bornean endemic *Arachnis calcarata* is an epiphyte (i.e., growing on other plants) perched on tall trees that reach to thirty metres in lower montane forests. Like many other epiphytes in tree crowns, it experiences an environment which is substantially drier than the forest floor and where nutrients are available only in small quantities in rain water. However, the thick hardy leaves of *Arachnis calcarata* are comparatively resistant to drying and, like most orchids, it produces aerial roots which are effective in absorbing moisture through the velamen, a spongy superficial layer of dead cells. The velamen usually gives a pale grey appearance to the aerial roots of orchids.

Arachnis orchids, like *Phalaenopsis* and *Aerides,* have a monopodial growth habit, in which the shoot tip grows continuously as the older parts below die and decay. *Arachnis calcarata* has a magnificent inflorescence that develops to about a metre long or more, and which is often longer than the whole plant itself. Its branched inflorescences produce many blooms at a time, each bloom up to five centimetres across and strikingly attractive owing to the maroon-brown spots and patches all over the yellow surfaces.

The species, with such large, long-lasting and sweetly smelling blooms and a hardy growth, has great potential in the breeding of ornamental scorpion orchids.

Plate 17

Pinanga salicifolia

The willow-leafed Pinang palm — Plate 18

One of the most attractive palms is *Pinanga salicifolia,* widespread in and endemic to Borneo. This *pinang* (as many species of *Pinanga* and *Areca* are called in Malay) often grows as clumps two or three metres tall, and is easily seen in peat swamp forests and some *kerangas* sites, and on steep ridgetops in Brunei. This graceful palm develops young shoots with narrow leaflets (recalling the slender leaves of the willow) but in the adult stems broad leaflets begin to form. Some mature stems continue to bear leaves with narrow leaflets and in this condition the palm is especially beautiful.

The species was first described in 1843, and in 1881 was cultivated in the Botanic Garden at Bogor in Java. Nowadays one seldom sees it, presumably because of a lack of introduction of new material for cultivation. It is certainly worthwhile cultivating as an indoor ornamental or in a shaded garden.

Palms are significant life forms in South East Asian rain forests. In all there are about 2,700 species of palm in the world, and about 280 species, or ten percent of the world total, in Borneo. Nearly half of all Bornean palms are now known to occur in Brunei.

Plate 18

Daemonorops formicaria

The devil's ant rattan — Plate 19

Rattans are common in the rain forest, and belong to the palm family. Most rattans are climbing and all are spiny plants producing fruit covered by overlapping scales.

Rattans are perhaps best known for their cane, used in the furniture industry, and about half a dozen species yield a cane of high commercial quality. Trade in rattan and cane furniture worldwide accounts for billions of dollars, and commercial exploitation of canes has led to dwindling natural populations of the most sought-after species. Consequently, supplier countries (all in South East Asia) are resorting to cultivating rattans as plantations and continuing the search for more rattans of comparable commercial quality.

One of the most interesting rattans in lowland and hill dipterocarp forest is a species of *Daemonorops,* which is a Greek name meaning "demon bush", probably coined in allusion to the ghastly spines on the leaf sheaths that tightly envelop the true stem or cane. This species, *Daemonorops formicaria,* forms an association with small black *Camponotus* ants just two or three millimeters in size (inset). On the stems of this rattan the ants build long, winding runways of interwoven fibres well protected by the rattan's spines. Rattans with such ant associations have been shown to absorb inorganic nutrients from the debris accumulated by these ants, hence deriving some benefit from their tenancy.

In Brunei this species is sometimes called *rotan bulu geok* which alludes to the resemblance of the soft bristly spines to the hairs of a caterpillar.

Plate 19

Bauhinia kockiana

A giant forest liana Plate 20

Of the huge woody climbers, or lianas, in the rain forest, one of the most memorable is surely *Bauhinia kockiana,* a vine of the bean family which reaches the highest tree crowns where they branch and leaf profusely. There, in the canopy of the forest, this vine comes into bloom, producing great masses of yellow and flaming red flowers easily noticed from a hill top or even a low-flying aircraft. Sometimes the liana is encountered along rivers where their tangles of branches and clusters of vividly coloured blooms decorate the riverside trees.

The genus takes its name from the famous Bauhin brothers, who were herbalists during the 16th century, as most species have leaves that are cleft at their tips or divided into identical halves. The plant climbs by clinging on with its coiled tendrils, produced along the younger branches.

Each flower has five stalked petals which are at first yellow and later become reddish orange. The dense clusters of flowers are extremely showy, and the species is occasionally cultivated as an ornamental in South East Asia. There are about 600 species of *Bauhinia* in the tropics, of which about 70 (mostly climbers) occur in Malesia, the botanical region stretching from Peninsular Thailand to New Guinea. Most cultivated species of *Bauhinia* are trees originating from India, South China and tropical America.

Plate 20

Merremia borneensis

The *Melabok* vine Plate 21

Merremia borneensis (*melabok* in Iban) belongs to the Convolvulaceae, a family that includes such familiar plants as the Morning Glory, the Sweet Potato and the *kangkong* (*Ipomoea aquatica*). This species is a woody vine common at forest fringes and in open sites on degraded soils, and is seldom encountered in primary forest. It is capable of prolific growth and can rapidly spread over an open earth slope, sometimes together with *Merremia peltata,* a sister species.

Merremia borneensis has not been cultivated as an ornamental in Borneo although its large, funnel-shaped yellow flowers are both pleasant and showy, probably because it is common along the wayside and is familiar as a plant of weedy growth. It should be remembered, however, that such plants grow easily even on the most difficult soil and that in a garden the vine would be maintained on a trellis or fence and subject to easy control. The vine flowers especially profusely during the drier months of February to April although blooms are available throughout the year.

The seed capsules ripen brown, each splitting into four parts to reveal four seeds. Fresh young fruits of *Merremia borneensis* were in the past used for preparing a hair tonic: the fruits were simply crushed and the sap applied directly to the hair.

Plate 21

Hornstedtia reticulata

The Spindle Ginger — Plate 22

Hornstedtia gingers are known as Spindle Gingers because they produce showy spindle-shaped flower heads formed by tightly overlapping, leathery purplish bracts. In each flowering head, the deep red tubular flowers are produced a few at a time, at the recessed topmost part where water is usually trapped. This has given rise to the name Drowning Men among some people, because the limp cleft flowers in the water that is trapped is likened to men waving their hands as they drown.

Hornstedtia reticulata (known as *telidus* by the Kedayan people, *sengang* in Iban and *sumbang* in Dusun) is a giant herb growing to three metres tall, with leaves about a metre long arranged in two rows along the erect stem (inset, right). The whole plant is often slightly raised above ground by thick stilt roots developing from the base. Each leaf is attached, by a constricted stalk-like portion, to a basal sheath that wraps tightly around the stem (inset, left). These sheaths are sometimes taken and dried, and strips are used for plaiting mats. The fruits are sweet and are occasionally eaten with rice.

The gingers as a group are largely aromatic herbs known for the various spices and condiments they supply, such as turmeric, cardamons and ginger itself. In the forest gingers are common in moist places and especially along rivers, in forest gaps and fringes. *Hornstedtia* is distributed from Peninsular Malaysia to New Guinea, the Solomon Islands and northern Queensland, and is represented in Borneo by eight species, of which four are endemic. *Hornstedtia reticulata* is a common lowland ginger and occurs frequently in alluvial forests and occasionally in peat swamp forests in Brunei. The sterile bracts on its inflorescence are the most strongly reticulated among the Bornean *Hornstedtia* species, and are rough to the touch.

Plate 22

Amydrium medium

A climbing *keladi* Plate 23

Many species of the taro family Araceae (*keladi* in Malay) abound in the mixed dipterocarp forests of Borneo. Among them are ground herbs, big and small, aquatic plants, climbers and epiphytes. Aroids, as they are also referred to, are characterised by their flowers which are closely packed all over a thick fleshy stalk, called a spadix (inset, right), the whole of which is more-or-less enclosed and protected by a leafy bract called a spathe. They include the common ornamentals of *Anthurium, Caladium, Dieffenbachia* and *Philodendron.*

Amydrium medium is an example of a climbing aroid that creeps up the trunks of forest trees using clasping roots produced along its stem. It is a common aroid in the lowland forests of Brunei, and belongs to the Monsteroid group of genera. The adult leaves of both *Amydrium medium* and *Monstera deliciosa,* a commonly cultivated ornamental, are perforated. The holes in the leaf blades of *Amydrium medium* are set close to the midrib, and do not develop in juvenile leaves (inset, left).

The spadix or inflorescence of *Amydrium* bears bisexual flowers. In some other aroids, the flowers are unisexual and are arranged on the spadix in discrete bands: females at the base, males at the tip, and a band of neutral (sterile) flowers in between, as in the cultivated taro plant *Colocasia esculenta.* The spathe in *Amydrium* is deciduous following floral maturity, although in certain aroids it may persist as a structure shielding the inflorescence.

Plate 23

Lecanopteris crustacea

An epiphytic ant-fern Plate 24

Lecanopteris crustacea is distributed in Borneo, Sumatra and Peninsular Malaysia. This unusual fern is an epiphyte growing in tree crowns of peat swamp forests and swampy spots beside rivers, and occasionally in the lowland mixed dipterocarp forests of Brunei. It has been found on the tall dipterocarps *Shorea pachyphylla* (in peat swamps) and *Dipterocarpus lowii* (in mixed dipterocarp forest), and occasionally associated with another epiphytic fern *Platycerium ridleyi.*

Lecanopteris is a genus of 15 species, all epiphytes and myrmecophytes, or ant-plants, housing ants in their fleshy, hollow rhizomes. The rhizome cavities form naturally, without excavation by the ants. In *Lecanopteris crustacea* the rhizomes are covered with tiny grey scales (inset) and form crust-like layers creeping over the host branch. The leaves grow in two rows on short conical projections along the rhizome and are of two types: sterile leaves whose function is photosynthesis, and longer fertile leaves with narrow lobes, which bear the spores on the lower surface.

In the association, the ants find a suitable habitat to breed in the hollows of the fern's rhizome, and in turn provide benefits to the plant through the organic material and detritus which they gather. The inner surface of the rhizome is dark-coloured, which not only provides for a suitable hiding place for ants but also may be important for the survival of symbiotic fungi that break down ant debris to release nutrients which the fern can absorb. The small, black ants in the rhizomes of *Lecanopteris crustacea* become greatly agitated and bite fiercely when the fern is disturbed, and may deter herbivores from the plant. These ants may also indirectly defend the fern by gleaning eggs and larvae of other insects from the leaves.

Plate 24

Ficus xylophylla

Strangling fig — Plate 25

Strangling figs begin life as epiphytes in tree crowns, and send down roots that gradually thicken on reaching the forest floor. These roots usually branch and self-graft on contact, forming a basket-like network around the trunk of the host tree. The crown of such a strangling fig begins to shade out that of its host tree, which sometimes dies more quickly as its bark is constricted by fig roots. Not all host trees die by constriction and sometimes the fig simply outlives its host.

A fig plant bears many tiny flowers on the inner surface of hollowed-out flowering heads called syconia or, simply, figs (inset, bottom right). The tip of each fig has an aperture protected by tightly overlapping scales (inset, top right), through which the specialised wasp pollinators of the fig can enter or leave. In the fig of *Ficus xylophylla* three kinds of flowers (diagram, bottom left) are found: male, producing the anthers and pollen (diagram: far left); female, bearing an ovary which produces the seed and a long style (diagram: far right); and gall, serving as an incubator for the wasp's larva (diagram: right-centre). Gall flowers have a style short enough for the wasp to insert its ovipositor during egg-laying and through which the emerging wasp will bite its way (diagram: left-centre).

This is the specialised setting for one of the most unusual adaptations between a plant and an animal. The wingless male wasps emerge from their gall flowers and, shortly after inseminating the female wasps, die within the fig. The winged female wasps then scramble out of the fig, in the process becoming laden with pollen from the fig's male flowers. They leave, and scramble into a young developing fig, where they attempt to lay eggs in the female and gall flowers. In so doing, they manage only to oviposit in gall flowers, which have styles short enough, and to pollinate female flowers which they fail to parasitize, following which they die. Hence, in this second fig, female flowers are pollinated and gall flowers are infected with wasp eggs. The maturation of new wasps is timed with the ripening of the fig.

Plate 25

Plate 26

Gigantochloa balui

A giant tree-grass Plate 26

Bamboos (called *buloh* in Malay) belong to the Grass family and are a distinct group characterised by their woody, hollow stems which impart rigidity and strength, and the much ramified branches produced at the base of each stem segment. The young shoots, known as *rebung* in Malay, have a specialised growth habit, where their elongation to achieve height precedes the development of the branches, so that new growth can proceed relatively unhampered in crowded clumps.

Among these giant tree-grasses, *Gigantochloa* bamboos include species that represent some of the most useful to man. The young shoots of some are relished as a delicacy, and the stems yield material for plaiting baskets and building walls, floors and other structures. In Brunei, Sabah and Sarawak, the well-known and commonly cultivated village bamboo called *buloh balui* (*Gigantochloa balui*) is such a bamboo. *Buloh balui* is also one of the most distinctive of *Gigantochloa* bamboos, and the protective sheaths that tightly embrace the young shoots are clothed with short silvery hairs that make the species distinctive.

Most *Gigantochloa* bamboos flower in grand fashion. The flowers of bamboos are produced in two rows on a short axis, together forming a specialised unit called a spikelet (inset, far left). Each flower is protected by an outer bract, visible in the intact spikelet, and an inner, more tender bract (inset, second from left). The anthers are held on filaments fused together in *Gigantochloa* to form a tube (inset, third from left) which surrounds the ovary and its style (inset, far right). The spikelets are produced in clusters along the branches, and eventually an entire bamboo stem, or even a whole clump, may bear such flowering branches, highly conspicuous from a distance. The entire flowering process of a clump from beginning to end can last well over a year, and the stems die after flowering. This can lead to the whole clump dying, but frequently it regenerates through new shoots produced from the subterranean rhizomes.

Gigantochloa bamboos prefer bright conditions and relatively fertile soils, and *buloh balui* thrives in open alluvial sites and beside large rivers in the lowlands. Botanists suspect that this species could have been introduced to Borneo a long time ago, as no indisputably wild populations have been discovered in Borneo whereas on the other hand what appears to be the same species has been noted as growing wild in parts of Thailand. In Java, some species of *Gigantochloa* are also not known to grow wild and are thought to reflect the past migrations of people who brought along their useful plants with them.

Resource and Heritage

As more forest disappears, it is vital to understand what might be lost when this great work of nature is decimated. In Brunei, the forests are both resource and heritage. An adequate area under forest ensures that the country does not develop at the expense of "ecological health". In a country with limited land, properly conserved forests guarantee a safe basic environment, and the continuing benefits of water and soil conservation. The wonderful array of life—both plant and animal—as well as the variety of habitat and landscape constitute Brunei's natural heritage. In addition to these services, the options and materials for future beneficial industries are conserved. Perhaps one instructive illustration is provided by the traditional multi-species, tree-dominated Malay garden, a heritage carried forward for generations which at the same time provides a small-scale diverse system from which a variety of useful products may be obtained. This is analogous to the rain forest, which is nature's multi-species garden from which we can continue to benefit, provided it is not destroyed. It is such a rich garden that we have not fully learnt the role and usefulness of every species within.

The world outlook for survival of tropical forests is generally one of great uncertainty. We are witnessing, as never before, a rate of loss of tropical forests due to the actions of man that can only be described as phenomenal. In the early 1980s, the estimated extent of tropical forest worldwide was 935 million hectares, eroded at an average rate of 24.5 million hectares a year, or 67,000 hectares a day. The situation has since become more dire, as loss of forest cover through commercial logging, slash and burn cultivation, rural fuelwood gathering and large-scale conversion to pasture land escalates.

The consequences of forest destruction on a grand scale such as this are frightening. Forests intercept rain water and gradually release it to supply rivers and streams without washing away too much soil. When the water and soil conservation roles of the forest are disturbed, it results in flash floods, dirty rivers, soil erosion and the silting of waterways, dams and coastal areas. Estimates in the 1970s revealed that flood damage in India incurred at an annual cost of US$140 to US$750 million. Argentina spends some US$10 million yearly to dredge silt from the estuary of the Plate River that has mainly washed down from the overgrazed catchment area of the Bermejo River 1,800 kilometres upstream, threatening to impair the value of Buenos Aires as a shipping port. The removal of forest cover, in which the bulk of nutrients are stored, leaves behind an impoverished soil where remaining nutrients may be easily washed away. It is known that heavy fertilizer inputs are required to raise a plantation crop on land from which forest had been cleared. With less accessible forest land, shifting cultivators in South East Asia have had to content themselves with shorter fallows during which the land cannot recover between periods of cultivation, with the result that poorer crop performances ensue, or moving to steeper terrain where more soil erosion is caused.

As forests decline in extent and the burning of fossil fuels increase worldwide, less carbon dioxide is converted back to organic carbon by plants and more carbon dioxide and pollutants are released. The balance is broken, and the carbon dioxide and particle content of the atmosphere increases, trapping more heat and leading to gradual global warming. If present trends continue, the earth's warmer temperatures will melt the polar ice caps sufficiently to raise sea levels, thereby flooding low-lying coastal areas, and cause shifts in the major climatic belts, affecting agricultural production areas. We are in an age where the whole of the environment is under seige. Overuse of poor soils in drought-prone areas turns millions of hectares into desert. Combustion of fossil fuels also results in acid rain, which can degrade significant tracts of forest. Pollution from industry and agriculture are harmful to people and wildlife. The proliferation of nuclear materials for both civil and military use increases the risk of accidents, as shown by the 1986 Chernobyl nuclear reactor explosion, which sent nuclear fallout across Europe. The World Commission on Environment and Development issued a strong statement in 1987, urging a rational and unified approach to strengthen waning options for the globe:

> *We have in the past been concerned about the impacts of economic growth upon the environment. We are now forced to concern ourselves with the impacts of ecological stress—degradation of soils, water regimes, atmosphere, and forests—upon our economic prospects. We have in the more recent past been forced to face up to a sharp increase in economic interdependence among nations. We are now forced to accustom ourselves to an accelerating ecological interdependence among nations.*

Marius Jacobs, the late conservationist, warned in 1983 before his death that "...man and nature have a common destiny, which reaches across thousands of years...only he is realistic who heeds the ecological realities."

A global concern

The message is exceedingly clear that the detrimental effects of environmental degradation have local, regional and global consequences, and do not observe national boundaries. While international cooperation may help in some areas, by and large the security of the global environment can only be achieved through the pursuit of ecologically sensible approaches to development within each region and by nations individually. The World Commission on Environment and Development observed: "The whole notion of security as traditionally understood...must be expanded to include the growing impacts of environmental stress... There are no military solutions to 'environmental insecurity'." What one country does will not only modify its own resources and environment but will have some effect on overall regional or global environmental stability; that final overall result will in turn affect the individual security of that country. One such global "evil loop" is outlined by the Commission: "Economic activity...could grow five- or tenfold...the world invests in houses, transport, farms and industries...the economic growth pulls raw material from forests, soils, seas, and waterways...new technology...also entails high risks, including new forms of pollution..." Pinpointing problems will help focus on weak areas but individual nations can no longer forge ahead without heeding the consequences of their policies and without due regard for "ecologically healthy" development.

Within this global context living systems are critical because their continued survival protects man and the physical environment, yet they are vulnerable to degradation by man and his manipulation of the physical environment. Living resources have a limited capacity to regenerate after damage. Taking "centre stage" are tropical rain forests, the richest ecosystem on earth and one which has rapidly declined in extent worldwide. The lessons are many and bitter. By 1983 the Philippines were estimated to have a remaining 38 percent forest cover and Thailand, 25 percent. Though small in land area, Brunei is greatly fortunate to still have some 80 percent forest cover, making rain forest the single largest and most visible living resource in the country.

Brunei's conservation areas

This has very special implications for Brunei. Its forests clothe the watershed areas that continue to sustain the major rivers with water. These are the steep uplands and hills where the main rivers Temburong, Temawai and Belalong (in Temburong District), Tutong (in Tutong District), Belait and Mendaram (in Belait District) and their tributaries originate, safe within the specially designated Protection Forests, Primary Conservation Areas and National Park. The latter two designations have been established to address the need for securing larger expanses of forest land which includes sufficient variation in natural landscape and vegetation types. The Batu Apoi forest area in the Temburong district, of 48,800 hectares, has been declared the Ulu Temburong National Park in 1991. This is Brunei's first national park, signifying a landmark achievement in adopting a comprehensive conservation system. The Primary Conservation Areas include the Sungai Ingei conservation area (18,500 hectares) and the Ulu Mendaram conservation area (6,200 hectares). They are in the main larger than the Protection Forests which were defined to conserve watersheds not within the Conservation Areas, and which are all within the Belait District (Bukit Teraja: 6,825 hectares; Bukit Batu Patam: 921 hectares; Bukit Ulu Tutong: 251 hectares) and Tutong District (Bukit Bedawan: 7,633 hectares; Benutan: 2,932 hectares). The Protection Forests should not be misconstrued as self-contained ecological units because they are not marked out as that; rather, they represent the core watershed areas which have been specially recognised and protected within reserved forest. The existence of such Protection Forests within the country should not be thought of as being the final seal of security for its water resources. This is because the river systems are maintained by the forest land abutting onto them as well; if these are unduly disturbed heavy surface erosion and siltation ruin the lower basins. For this reason the forests of the lowlands are of crucial importance as well.

One case in point is the large peat swamp complex of the Mendaram and Belait areas. Peat swamps are not giant compost heaps of rotting, wet debris. They are the specialised ecosystems safeguarding coastal climatic and hydrological regimes which exist in balance with the rivers and streams around, and destruction of these swamp systems are likely to adversely affect the quality of rivers, streams and the coastal areas. In other countries the acidity of swampland has been a major constraint to instituting most forms of agriculture, which is essentially a poor substitute for the services which swamps provide in their hydrological and protective roles. Larger countries have often unflinchingly sacrificed significant portions of their swamp forests, but there is no case for a country of limited land area, such as Brunei, to gamble. Once a swamp forest is filled and converted, it is gone forever. In addition, the Brunei peat swamp forests are among the tallest swamp forests in the world, and are characterised by many special aspects of vegetation and ecology, as we have seen in earlier discussions. In the face of widespread conversion and degradation of peat swamp forests in many parts of South East Asia, the ones existing in Brunei begin to hold an even more special position. In New Zealand, the loss of wild kauri pines *(Agathis)* are mourned by foresters there and visitors are even brought on special guided tours to admire these once abundantly wild trees. In Brunei, in the centres of the peat swamps, accessible groves of wild *Agathis* still exist. Surely the value of these trees as seed sources and a special facet of Brunei's natural history, coupled with the structural grandeur and ecological significance of the peat swamp forests in general, constitute a very special heritage. Perhaps the vulnerability of peat swamps were most highlighted by the extensive forest fires of 1983, which ravaged huge areas of this resource in Kalimantan. Even before this, peat swamp forest fires following periods of severe drought have been known in the South Sumatra and Riau provinces of Indonesia, in the Saribas swamps in Sarawak and in

Brunei, where extensive damage was inflicted on the Anduki Forest Reserve in 1958. Whereas in Sumatra the ecologist Rob Anderson noted that the fire left behind "a waterlogged morass in which nothing, except water plants, will grow", the Anduki peat swamp forest in Brunei has not been totally lost. Today the route to Seria passes by dense, actively regenerating stands of the swamp dipterocarp *kapur paya* (*Dryobalanops rappa*). These stands not only have aesthetic value along the journey to Seria, but are important to forestry in that they are almost pure stands that can be studied from the viewpoint of regeneration following fire destruction and as a source of seeds for the future. Pure stands of *Dryobalanops rappa* nowadays rank among the world's rarest sights and the ones in Anduki Forest Reserve (about 900 hectares) are now kept as a Secondary Conservation Area. Conservation has much to do with the future and in time the value of these stands can only grow.

Besides the Anduki forest, several more sites have been designated Secondary Conservation Areas, set up to preserve the integrity of smaller units of particular forest types. These include the Badas Forest Reserve, in which a high concentration of the conifer *Agathis* is to be found in an area only about 100 hectares in extent, and the twin compartments 7 and 8 of the Andulau Forest Reserve (1,300 hectares), conserving an undisturbed patch of coastal lowland mixed dipterocarp forest where some of the earliest ecological studies of dipterocarp rain forest were conducted by Ashton, and which is now used for biological research by the Forestry Department and the University of Brunei Darussalam. In addition, the Peradayan Forest Reserve (about 1,000 hectares) which occupies a sandstone ridge surmounted by hill *kerangas* and skirted by mixed dipterocarp forest, is also kept as a Secondary Conservation area, a small portion of which is open to public recreation.

These are the main conservation forest sites of Brunei and recent plans propose to add more conservation areas in mangrove, peat and fresh water swamp areas, and in the coastal *kerangas* sites which are being threatened with degradation in the whole of Borneo. Coastal resources are particularly vulnerable and their destruction often excused because they are located where the major population centres are developing. However, these very centres of human settlement will become physically and climatically affected sites once the entire coastal belt is deforested.

Human populations keep growing, but land area and resources do not expand. Ecologically sensible planning of resource utilisation and conservation must be given priority. Ali Moertopo, a Minister of Information for Indonesia, one of the tropics' most populous countries, once summed up the situation when he noted that "we should also conceive threat in terms of our inability to maintain a new balance between the carrying capacities of the earth's land area...and the fast rate of population growth. If the balance is not redressed...sooner or later, the world will perish."

The rich lowland forests

The lowland rain forests are by far the richest biologically among the different types. While the huge socio-economic benefits that these forests grant in their soil and water conservation roles are often taken for granted, their usefulness as a source of timber, and more recently their degradation because of this, are widely realised. A study in Kalimantan revealed that the removal of one big tree during forest logging destroyed seventeen others and damaged nearly half of the crowns of the remaining ones. Another in Peninsular Malaysia found that one extraction removing approximately ten percent of the standing stock, of trees ten centimetres or greater in diameter, destroyed another 55 percent, leaving only 35 percent undamaged. Ecological webs that maintain the integrity of these forests are disrupted. The top soil, in which nutrient-cycling micro-organisms survive, is often disturbed and even removed in the logging tracks criss-crossing the forest. In intensive logging, the best and tallest trees are often removed, leaving behind a poorer stand where patches are modified to such an extent by the increased light and evaporation that many species fail to regenerate. At the same time there is a reduction in seed and fruit supplies and animal populations are affected. Even the so-called "selective logging" procedures, which remove a prescribed quantity of stems of certain species, will change the forest after several cycles of cutting because the slower growing species would not have reached reproductive maturity before they are disturbed again. After the dominant trees are artificially removed by logging, the regenerating forest is likely not to attain the original height, because height-growth is not encouraged by the resulting, more open conditions.

This reduces the volume of living space of the forest framework, which originally provided the maximum space opportunities to which different plants and animals are naturally adapted. Silvicultural operations, with the best of intentions in carrying out forestry operations to promote regrowth of logged-over forests through such activities as "enrichment planting" of choice species, simply cannot keep up with the high rates of logging in some countries. Estimates in 1983 indicated that annual loss of forest cover to commercial logging amounted to between 0.8 million and 2.5 million hectares in Latin America, 3.2 million hectares in tropical Africa and between 1 million and 2.7 million hectares in South East Asia.

We stand to lose many species from the rich lowland forests if logging on a wide scale is permitted. Many larger animals are widely wandering and require a large area of forest to survive, and birds are documented to be severely disturbed by logging. The orang utan of Borneo and Sumatra have been found to be absent or severely depleted in numbers following logging. Plants which are highly localised, or even endemic to a small area, such as the shade-adapted dwarf fan-palm *Livistona exigua* (known from only a few localities in Brunei), can be easily extinguished by forest disturbance through logging. Other plants, especially many wild fruit trees, occur at very low densities in the rain forest and some are even dioecious (with male and female flowers on separate trees); therefore such plants can only breed effectively within large tracts of forest land containing enough individuals and where their specialised pollinators and dispersers are found. According to one study of a hectare of forest in Temburong during 1991–92, as many as 143 of 231 species of tree (i.e., more than half of all tree species) were represented by single individuals. For a dioecious rain forest tree, it has been estimated that at least five hundred mature individuals within the same forest area might be representative of a breeding population where different variable traits are adequately represented. The most common of these trees may occur in densities of perhaps two dozen individuals in a hundred hectares of forest, so that at least 2,000 hectares may contain five hundred breeding individuals. But the complement of species from one place to another is not the same, because these forests are so rich and the sites where each species can establish are not uniformly distributed. Summarising, unless the forests are maintained in large tracts, many species or populations will be lost.

A multitude of products

The maintenance of biodiversity must continue to be a primary function of the forest areas of Brunei. Many products other than timber have been derived from forests. One of the earliest forest products of international importance in the Far East was Borneo Camphor, a resinous exudate of several *kapur* (*Dryobalanops*) species, which was already being actively traded by Arabs during the 6th Century. So important was this product that the 13th-century Chinese work *Chu-fan Chi* (written around 1225 A.D.) referred to the grades of camphor originating from Brunei and Palembang in Sumatra. Pigafetta, Magellan's chronicler during his voyage around the world, also described in 1521 a lively trade in forest and vegetable products, including camphor, centred in Brunei. Camphor has now waned in importance but many other forest products are now acknowledged in South East Asia as being commercially important or playing a significant role in the household or rural economy. Trade in rattan and rattan products from South East Asian forests, including furniture, has assumed worldwide importance. Bamboo is important in rural construction and also, with rattan, *Pandanus* leaves (*mengkuang*) and *Donax* stems, for local handicrafts. The modification of traditional handcrafted items to suit modern situations, such as decorative screens plaited from natural materials, is boosting the commercial prospects of handicrafts. Bamboo shoots (*rebung*) and palm cabbages (*umbut*) are locally varied and popular, as are the numerous types of fruits and nuts. The *Nypa* palm, abundant in estuarine areas, has for a long time been used as a source of thatch, palm sugar and fruit. Bumper harvests of illipe nuts are available only every few years and even then is marketed overseas for use in the cosmetic and confectionery industries. Sago starch and medicinal tonics are important in South East Asia. A tonic prepared from *Eurycoma longifolia* (*tongkat ali*), used to promote general health and male vitality, is even now available commercially. Honey from the bee *Apis dorsata,* which in Borneo commonly builds its colonies on the tall *menggaris* (*Koompassia excelsa*), is used and traded in the region. The list grows longer, but the point is that a great many products already useful to people are from forests.

Growing prospects for several categories of forest products are now actively pursued by many countries. Plantations of rattans are now being set up in several countries to supply cane for the future. Commercially important cane species number perhaps only a dozen, but the rich forest base may contain further choice rattans that can be exploited. The ornamental plant and horticultural industries stand to benefit most from the forest's biodiversity, because the breeding and selection of new or improved plants can only be carried out if in the first place a variety of species or forms is available. The orchid industry will continue to depend on wild species in order to generate new and better hybrids. *Ixora* and *Rhododendron* are two groups of ornamental plants for which wild relatives are similarly important, and pitcher plants (*Nepenthes*) are gaining popularity, though as yet not specially bred. Palms and flowering trees and ferns, too, have immense commercial potential as ornamental plants. Fruit-tree horticulture will in the same way benefit from the existence of wild relatives in the forest. Although Alfred Russel Wallace said in the 19th century that "to eat Durians is a new sensation, worth a voyage to the East to experience," there are now many improved varieties. In fact, the common *durian* (*Durio zibethinus*) has been so much bred that it is known only in cultivation. Yet there are many species of wild *durian* (including *Durio dulcis, Durio graveolens, Durio oxleyanus* and *Durio kutejensis* in Brunei) said to have good or even better fruits, which can be tried and improved by horticulture. Perhaps it may even be possible to exploit wild *rambutan* (*Nephelium*) species adapted to nutrient-poor soils such as in *kerangas* in order to produce varieties or grafts that are more hardy. Mangoes (*Mangifera*) are also represented by forest species providing a large variety of fruit size, texture, taste and productivity. There is every reason to expect that with proper selection the Bornean *membangan* mango (*Mangifera pajang*) can be as commercially popular as the Indian mango (*Mangifera indica*).

Nowadays tropical rain forests are often referred to as a "pharmaceutical storehouse" due to the range of drugs and medicines that can be prepared from many wild forest plants. The principal compounds involved are defence substances produced by plants to repel herbivores, from which they cannot run. Some of these chemicals (called secondary compounds because they have no essential role in the plant's normal metabolism) are toxic but many merely reduce the palatability of the tissues to herbivores. Some possess medicinal properties and are well known to Southeast Asian cultures that have long traditions of indigenous medicine. Thus people in Brunei have inherited folk notions that the petals of a *Melastoma* flower are effective when rubbed onto the skin to remove scars of superficial wounds, and that shavings of branch wood (called *raja kayu*) from the *tulong* tree (*Agathis borneensis*) when boiled with water provides a drink effective in treating high blood pressure and stomach aches. Much of such traditional uses remain to be understood in biochemical terms but many are already well investigated. Extracts of the vine *Tinospora crispa,* much used for treating bacterial abscess, have been shown to have bactericidal properties against *Staphylococcus* and *Streptococcus*. In Thailand, roots of *Rauwolfia serpentina,* which contain reserpine and rescinamine (commercially used as hypotensives, in treating high blood pressure, and sedatives), are used locally and supplied to international pharmaceutical companies. The chemicals from plants are not restricted to medicinal uses and are important also as tanning agents, stabilizers and dyes. The roots of *Derris elliptica,* a forest vine, yield an abstract much used in Borneo as a fish poison and as an insecticide and in fact contains rotenone, a commercially important insecticide. Bark of the *kelimpanas* (*Goniothalamus*) is widely and effectively used in rural areas in Brunei and Sarawak by burning to repel insects.

Plant compounds have great value in many ways and are either used directly, or as raw materials in the partial synthesis of drugs. These compounds also serve as useful models in the modification and improvement of pharmaceutical products. In spite of a resurgence of interest in folk medicine and the chemical screening of plants, only a tiny fraction of rain-forest species have been investigated. The traditional pharmacopoeia of South East Asia has itself been estimated to have only utilised less than five percent of the indigenous flora, which in turn is far from fully documented. We are confronted with the choice of keeping all these options open, or losing everything by allowing the destruction and degradation of the forests.

Multiple-use resource management

A multi-species system cannot be regarded as a single entity yielding only one or two values. The traditional Malay home garden reflects this in its make-up of a variety of plants, chiefly fruit trees, yielding a range of useful products available continuously or at different seasons. Different types of fruit trees are raised, some mature and others newly planted, but all within a system regarded as essentially long-term and passed on to the following generation. The home garden will contain such plants as bamboos, palms and bananas, but the variety of trees gives it its distinctive character and also shields the home from the elements. In a way this has been an ecologically sensible model in an environment that is characterised by species-rich forests. Now, as modern development proceeds and urban centres are built up, it is important that the rain forest is regarded as nature's multi-species garden. For not only does the rain forest protect our soils, climate, water supply, river banks and coastlines, they are also the breeding grounds for numerous types of plants and animals and the source of an unrivalled variety of products. For modern man its destruction is and will be a source of misery and risk. Within the limited land area of Brunei, there is only good that can come from sustaining the rain forest and exploiting parts of it in a non-destructive sustainable manner.

These forests are the living monuments of nature which form part of the great heritage of Brunei. They have been, and will continue to be the subject of great fascination, admiration, and the basis of scientific advancement. It will largely be due to their existence that Brunei continues to be environmentally secure and ecologically healthy. If this resource is to continue to be a heritage of the nation, its survival and continuing functions must be guaranteed. And this heritage can only at the same time be a resource if we use and conserve it wisely, keeping options open for our future.

Garcinia mangostana

The "Queen of Fruits" Plate 27

The mangosteen has been hailed as the "Queen of Fruits" and "the finest fruit in the world." Jacobus Bontius, writing on the virtues of the mangosteen, remarked that it was "of all the fruits of the Indies by far the most delicious." This is not surprising, because within the purple rind of the fruit, which is only as large as a tennis ball, the pearly white flesh of the seeds has a delicate, sweet flavour highly agreeable to the palates of widely different cultures. In Brunei, Indonesia and Malaysia, the mangosteen is popularly known as *manggis.*

For all its popularity in Asia and even in Europe, the mangosteen is widely cultivated only in the Far East, although there are orchards in Madagascar, Queensland, the Honduras and Brazil. This is perhaps because mangosteen seedlings are notoriously difficult to raise and require fertile, well-drained soils. Until today no truly effective methods for mass-propagating the mangosteen have been found, and the fruit remains as much an enigma as it is the "prize of the Indies". Wild trees of the mangosteen are either rare or non-existent, and the fruits form without fertilisation (the embryo develops from the tissue of the inner ovary wall), so that the seedlings are true to type, i.e., they are genetic clones of the mother tree. This lack of variation makes it difficult to select for improved qualities in breeding. It has also been suggested that the mangosteen may have originated as a hybrid of the wild species *Garcinia hombroniana* and *Garcinia malaccensis.*

The mangosteen genus is a large one, of about 400 species, including a number of other useful species. Gambodge is a yellow pigment obtained from the latex of an Indo-Chinese species, after which Cambodia is named, as well as several other species. The trees are tapped to obtain gambodge used in water colours and dyeing, and the robes of Buddhist priests were traditionally dyed with this. The *asam gelugor* is *Garcinia atroviridis,* wild in Burma and Peninsular Malaysia, of which the dried fruit rind is used as a sour seasoning for food. The tannin-rich rind of the mangosteen has been used to tan leather, in traditional treatment of dysentery and as a substance to fix black dyes. The seeds of *Garcinia*

Plate 27

indica from tropical Asia yield an edible fat called *kokam* or Goa butter, and *Garcinia pedunculata* from Bengal produces a large edible fruit.

The mangosteen renders a special lesson. Though it must have been cultivated since long ago, the poor chances of obtaining wild material means we have to be content with the fruit as it is, though of course the taste has been widely appreciated. For many of our other fruit trees—including the *rambutan, langsat,* mango, *terap* and *chempedak*—wild trees and wild relatives abound in the rain forest. Only by conserving these genetic resources can we retain the chance to exploit the best qualities for improving the crops we desire.

Rhododendron longiflorum

Blooms for the future? Plate 28

Rhododendron, derived from the Greek words meaning "rose trees", is the name of the horticulturally important genus of the European alpenrose. Most species have showy, brightly coloured blooms. The genus has about 900 species, mostly distributed in the Himalayan region and Nepal to west China, but found also in North America, Europe and the Orient; there is only a single species in Australia.

In Borneo there are about fifty species, of which a dozen or so occur in Brunei. Most of the species occur in montane forests or as epiphytes along rivers in the lowlands. *Rhododendron longiflorum* is exceptional in being a lowland species growing as a ground shrub and therefore has potential as an ornamental for the tropical lowlands. It was first introduced into cultivation in European hot houses around 1860 and has been used for the breeding of many hybrids. In the 1930s Professor Holttum of the Singapore Botanic Gardens attempted hybridisation work with this species, though unfortunately the trials came to an end with the Second World War. The clusters of large, orange-pink flowers and the freely flowering habit suggest that further horticultural trials with this species are extremely worthwhile. The plant illustrated here was found growing on the harsh soils of some *kerangas* sites and so far this form of the species has only been documented in Brunei and southwest Sabah. Its adaptation to nutrient-poor sandy soils also implies it will be hardy in cultivation.

Rhododendron flowers are protandrous, meaning the male parts (anthers) of a flower mature before the female part (the stigma). When the flower first opens the anthers are already ripe and release masses of pollen, while the stigma remains quite dry. Later the style elongates and the stigma becomes sticky and receptive to pollen from other flowers. This habit appears to ensure that cross-pollination between individuals takes place. The tiny seeds are light and dispersed by wind following their release from ruptured ripe fruits.

Plate 28

Arachnis hookeriana

The ornamental wonder   Plate 29

The scorpion orchids, *Arachnis,* like orchids of the well-known genera *Dendrobium, Phalaenopsis* and *Vanda*, are of great commercial importance to the horticulture industry. *Arachnis hookeriana* can be said to be the most important of scorpion orchids because it is very much used in the production of hybrid scorpion orchids, well liked for the pretty, long-lived blooms produced in abundance. Several hybrids are not only garden plants but are also used in the mass-production of cut-flowers.

In the breeding of plants, two situations need to be carefully considered. Once the correct mixture of characteristics is obtained in a hybrid or stock, it may be necessary to mass-produce plants in which a constant quality is assured. On the other hand, there is a limit to the variety that can be obtained with only several parents, and constant inbreeding can result in undesirable traits in a plant. This means that fresh genetic stock from wild plants of different provenances (populations or localities) is often necessary for the continued production of new varieties and better stocks.

Thus, wild populations of plants such as *Arachnis hookeriana* continue to be important. Not only should their value be realised, but these populations must also be safeguarded from total decimation by thoughtless collecting and habitat destruction. *Arachnis hookeriana* is a terrestrial scrambling orchid not uncommon in the lowlands of Peninsular Malaysia and Borneo, including Brunei Darussalam. This seemingly unspectacular orchid represents one of the most important plants in the horticultural industry in the region!

Plate 29

Alocasia lowii

Maroon-leafed forest *keladi* Plate 30

This member of the taro family, Araceae (*keladi* in Malay), is an attractive herb of the forest floor in mixed dipterocarp forest, particularly in alluvial areas. The maroon-mottled leaf stalks, which have probably prompted the name *keladi lanut* ("snake *keladi*" in Dusun), and the frequently maroon lower leaf surface, are very distinctive of this species, and give it ornamental value.

As a forest-floor herb of shaded, sometimes rather dark, habitats, *Alocasia lowii* survives in conditions where not only is overall light intensity reduced, but the quantity of light energy for photosynthesis is lowered through the filtering effect of the forest canopy. Species such as *Alocasia lowii* can develop anthocyanin (which causes the maroon colouration) on the lower side of the leaf, which reflects light energy back through the leaf's photosynthetic tissue. This back-scattering of light that would otherwise have passed through the leaf improves the efficiency of energy capture, allowing the plant to survive under diminished light conditions.

Many plants accumulate special chemical compounds, mostly metabolic by-products, that appear to enhance their survival in the competitive environment of the rain forest. Whereas some of these so-called secondary compounds assist in enhancing light-capture, others render plant tissues unpalatable to herbivores. Some induce biochemical effects in animal tissue and a number are medicinally useful to humans. From this perspective, the rain forest can be regarded as a pharmaceutical storehouse, where many species manufacture compounds that are useful or potentially useful drugs.

Plate 30

Index to Plant Scientific Names

(page numbers in bold indicate illustrations)

Index to Plant Common Names

(page numbers in bold indicate illustrations)

General Index